An OPUS book

THE PROBLEMS OF PHYSICS

OPUS General Editors

Keith Thomas
Alan Ryan
Walter Bodmer

OPUS books provide concise, original, and authoritative introductions to a wide range of subjects in the humanities and sciences. They are written by experts for the general reader as well as for students.

The Problems of Science

This group of OPUS books describes the current state of key scientific subjects, with special emphasis on the questions now at the forefront of research.

The Problems of Physics A. J. Leggett
The Problems of Biology John Maynard Smith
The Problems of Chemistry W. Graham Richards
The Problems of Evolution Mark Ridley
The Problems of Mathematics Ian Stewart

The Problems of Physics

A. J. LEGGETT

Oxford New York
OXFORD UNIVERSITY PRESS
1987

Oxford University Press, Walton Street, Oxford OX2 6DP
Oxford New York Toronto
Delhi Bombay Calcutta Madras Karachi
Petaling Jaya Singapore Hong Kong Tokyo
Nairobi Dar es Salaam Cape Town
Melbourne Auckland
and associated companies in
Beirut Berlin Ibadan Nicosia

Oxford is a trade mark of Oxford University Press

First published 1987 as an Oxford University Press paperback
and simultaneously in a hardback edition

British Library Cataloguing in Publication Data
Leggett, A. J.
The problems of physics. — (OPUS).
1. Physics
I. Title
530 QC21.2
ISBN 0–19–219205–1
ISBN 0–19–289186–3 Pbk

Library of Congress Cataloging in Publication Data
Leggett, A. J.
The problems of physics.
(OPUS)
1. Physics. 2. Particles (Nuclear physics)
3. Condensed matter. 4. Cosmology.
I. Title. II. Series.
QC21.2.L44 1987 530 87–12278
ISBN 0–19–219205–1
ISBN 0–19–289186–3 (pbk.)

Set by Dobbie Typesetting Service
Printed and bound in Great Britain by
The Guernsey Press Co. Ltd, Guernsey, Channel Islands

To my parents.
To Haruko and Asako.

Preface

This book is intended as an introduction to some of the major problems studied by contemporary physicists. I have assumed that the reader is familiar with those aspects of the atomic picture of matter which can reasonably be said to be part of our contemporary culture but, apart from this, does not necessarily have any technical background. Where arguments about specific points of physics are given in the text, I have tried to make them self-contained.

It would be quite impossible in a book of this length to give even a superficial discussion of all the diverse aspects and areas of the subject we call 'physics'. I have had, perforce, to ignore totally not only the 'applied', or technological, aspects of the subject, but also a whole range of fascinating questions concerning its organization and sociology. Even within physics regarded purely as an academic discipline, I have not attempted complete coverage; huge subfields such as atomic, molecular, and even nuclear physics are completely unrepresented in this book, and others such as astrophysics and biophysics are mentioned only briefly. Rather, after an introduction which attempts to review how we got to where we are, I have concentrated on four major 'frontier' areas of current research which I believe are reasonably representative of the subject: particle physics, cosmology, condensed-matter physics, and 'foundational' questions, the frontiers corresponding, one might say, to the very small, the very large, the very complex, and the very unclear. I do not believe that inclusion of other subfields such as geophysics or nuclear physics would introduce many *qualitatively* new features which are not already exemplified in these four areas.

The focus of this book is the current *problems* of physics, not the answers which physics has already provided. Thus, I have spent time detailing our current picture only to the extent that this is necessary in order to attain a vantage-point from which the problems can be viewed. Also, while I have tried to explain schematically the basic principles involved in the acquisition of experimental information in the various areas, I have not attempted to discuss the practical aspects of experimental design. The reader who wishes for more detailed information, on either the structure of existing theory or the experimental methods currently in use, may consult the various books and articles listed under 'further reading'.

Finally, a word to my professional colleagues, in case any of them should happen to pick up this book: it is not meant for them! They will undoubtedly find some points at which they will disagree with my presentation; indeed, I have repeatedly and consciously had to resist the temptation to qualify some of my statements with the string of technical reservations which I am well aware would be essential in a journal review article, say, but which would make the book quite indigestible to a non-specialist readership. In particular, I am acutely conscious that the brief discussion in the last chapter of the evidence—or rather lack of it—for the applicability of the quantum-mechanical formalism *in toto* to complex systems grossly oversimplifies a highly complex and technical issue, and that, read in isolation, it may seem to some professional physicists misleading and possibly even outrageous. All I can say is that I have given extended and technically detailed discussions of this topic elsewhere; and I would appeal to potential critics, before they indignantly take pen in hand to tell me how self-evidently ill-informed and preposterous my assertions are, to read also these more technical papers.

I am grateful to many colleagues at the University of Illinois and elsewhere for discussions which have helped me to clarify my thoughts on some of the subjects discussed here; in particular, I thank Gordon Baym, Mike Stone, Jon Thaler, Bill Watson, and Bill Wyld for reading parts of the draft manuscript and commenting on it. In a more general way, the overall attitude to the established canon of physics which is implicit in this book has been influenced by conversations over many years with Brian Easlea and Aaron Sloman; to them also I extend my thanks. Finally, I am grateful to my wife and daughter for their moral support, and for putting up with the disruption of domestic life entailed by the writing of this book.

A. J. LEGGETT

Contents

List of figures

1
Setting the stage

The word 'physics' goes back to Aristotle, who wrote a book with that title. But the derivation is from the Greek word *physis*, meaning 'growth' or 'nature'; and many of the questions which Aristotle discussed would today be more naturally classified as the subject-matter of philosophy or biology. Indeed, the name did not stick, and for many centuries the subject we now call 'physics' was called 'natural philosophy'—that is, philosophy of nature. In Britain, some professors of physics still hold chairs with this title. Let us start by taking a quick look at a few of the milestones in the development of the subject over the last few hundred years. In the process I will try to point out to the reader some of the essential concepts we will meet in later chapters.

Although some isolated elements of the world-view enshrined in modern physics can certainly be traced back to the cultures of China, India, Greece, and other civilizations more than two thousand years ago, the origins of the subject as the coherent, quantitative discipline we now know would probably be placed by most historians in Europe in the late medieval and early Renaissance period. Just how and why these seminal developments occurred when and where they did, how they reflected much older traditions in the culture, and how some of these pre-scientific traditions may still be colouring the conscious or unconscious assumptions of physicists today—all these are fascinating questions, but I have neither the space nor the historical expertise to discuss them in this book. In the present context, it is enough to recall that physics as we know it began with the systematic and quantitative study of two apparently disparate subjects: mechanics and astronomy. In each case the technological developments of the period were essential ingredients in their development. This is obvious in the case of astronomy, where the invention of the telescope around 1600 was a major landmark; but in mechanics, the development of accurate clocks played an even more fundamental role, perhaps. For the first time a reproducible and quantitative standard not only of length but of time was available, and one could begin to think quantitatively about concepts such as velocity and acceleration, which would eventually form the language in which Newton's dynamics was formulated. As the technology of the sixteenth and seventeenth centuries developed, more and more of the concepts of macroscopic physics as we now know it—mass, force, pressure, temperature, and so on—began to acquire a quantitative meaning, and many empirical laws relating to them were found, some of which are well known to every schoolchild today: Snell's law for the refraction of light in a material such as glass, Boyle's law relating the pressure and the volume of a gas, Hooke's law relating the extension of an elastic spring to the force applied to it, and so on. At the same time, in the sister subject of chemistry, scientists began to formulate quantitatively the laws according to which different substances combine, although at the time—and indeed for a couple of centuries or so thereafter—it was not clear what these generalizations had to do with the mechanical problems investigated by the physicists.

The man who is usually regarded as the father of physics as we know it is of course Isaac Newton (1642–1727). Newton made fundamental contributions to many different branches of physics, but it is his work on mechanics and astronomy which has left an indelible stamp on the subject. What he did was, first, to formulate explicitly the basic laws of the mechanics of macroscopic bodies which we still believe to be valid today (to the extent that the effects of quantum mechanics and special and general relativity can be neglected, which is usually an excellent approximation at least for terrestrial bodies); second, to develop with others the relevant mathematics (differential and integral calculus) to the point where the equations of motion could actually be solved for a number of interesting physical situations; and third, to recognize that the principles of mechanics were equally valid on earth and in the heavens, and specifically, that the force which held the planets in their orbits around the sun was the very same force, gravitation, which was responsible for the downward pull on objects on earth. In each case, the gravitational force was proportional to the product of the masses of the bodies involved and inversely proportional to the square of their separation.

Newton's laws of motion are so fundamental a bedrock of modern physics that it is difficult to imagine what the subject would have been like without them. It is worth giving them explicitly here:

(1) Every body continues in its state of rest, or of uniform motion in a straight line, unless it is compelled to change that state by forces impressed upon it.

(2) The change of motion is proportional to the motive force impressed, and occurs in the direction of the straight line in which that force is impressed (that is, it is parallel to the force).

(3) To every action is always opposed an equal reaction; or, the mutual actions of two bodies upon each other are always equal, and directed to contrary parts (that is, in opposite directions).

Perhaps even more fundamental than the detailed form of the laws themselves are the assumptions implicit in them about the kinds of questions we want to ask. Consider in particular the second law, which in modern notation reads: the acceleration of a body is equal to the force acting on it divided by its mass. Acceleration is the rate of change of velocity with time, and velocity is the change of position with time; so, if the force on a body as well as that body's mass are known, we can calculate the 'rate of change of the rate of change' of the position with time. In mathematical terms, the equation of motion is a second-order differential equation with respect to time; so, to obtain a definite solution, we need two additional pieces of information: for example, the position of the body at some initial time and its velocity at that time. As a matter of fact, these are not the only items of knowledge which uniquely fix the solution: for example, if we specify the position of the body at both the initial and the final time, the solution is again uniquely determined. However, it is fairly obvious that in many practical problems—for example, in trying to calculate the trajectory of a cannon-ball or the motion of the planets—the two pieces of information which we are most likely to have are indeed the initial position and velocity. Consequently, we tend to look upon the process of specifying these initial data and then using Newton's second law to derive the behaviour of the body in question at a later time as a paradigm of 'explanation' in mechanics; and then, by extrapolation, to assume that all or most explanation in the physical sciences—all explanation of specific events at least—must consist in a procedure which will permit us, when given the initial state of a system, to predict or derive its subsequent behaviour. For example, in cosmology we tend to assume without even thinking about it that an 'explanation' of the present state of the universe must refer to its past: the universe is the way it is 'because of' the way it started, not because of the way it will end or for any other type of reason. It is interesting to reflect that this view of 'explanation', which is so clearly rooted in purely anthropomorphic considerations, has withstood the violent conceptual revolutions of relativity and quantum mechanics, which, one might think, would make these considerations quite irrelevant much of the time.

Newton's formulation of the basic equations of classical mechanics would have made much less impact, of course, in the absence of his second great contribution, the development of the mathematics necessary to solve them in interesting situations. However, his third achievement—the recognition of the unity of the laws of mechanics on earth and in the heavens—was perhaps, in the long run, the most important of all. Today it is difficult for us to appreciate the colossal conceptual leap involved in the postulate that the force which made the anecdotal apple fall to the ground was the very same force which held the planets in their orbits. This is the first in a long line of great 'unifications' in the history of physics, in which apparently totally disparate phenomena of nature were recognized as different manifestations of the same basic effect. Here, as in some other cases, the perception of unity involved extrapolation far beyond what one could measure directly: Newton could devise methods for measuring the gravitational force directly on earth, but he had to postulate that this force acted on the scale of the solar system, and verify this by working out the consequences, some of which were to become *directly* testable only centuries later. At first sight, the lesson of the colossal success of Newton's mechanics is that it pays to be bold in extrapolating the laws of nature from the comparatively small region in which we can test them directly to regions vastly distant, not just in space or time but, for example, in the density of matter involved—and, as we shall see, in some areas of physics that lesson has been very well learned indeed.

For the century or so after Newton's death, the course of physics ran comparatively smoothly. On the one hand, great progress was made in applying his laws of motion to more and more complicated problems in mechanics. To a large extent, this development could be regarded as lying in the area of applied mathematics, rather than physics, since the problem was simply to solve Newton's equations for more and more complicated situations; and once the initial state and the forces are specified, this is a purely mathematical operation, even though its 'pay-off' lies in the area of physics. (I will return to this point in more generality below.) In the course of this mathematical development, many elegant reformulations of Newton's laws were devised (for example, one well-known such principle states that a body will follow that path between two points which, given certain conditions, takes the shortest time); and it eventually turned out that some of these alternative formulations were a crucial clue in the development of quantum mechanics a century or more later. It is an interesting thought that, had mathematicians of the late eighteenth and early nineteenth centuries had access to modern computing facilities, they would have been deprived of much of the motivation for developing these elegant reformulations, and the eventual transition to quantum mechanics would very likely have been even more traumatic than it in fact was.

In a different direction, a great deal of progress was made, both experimentally and theoretically, in this period in the areas of electricity and magnetism, optics and thermodynamics—what we today think of as the subject-matter of 'classical' physics. From a study of static electrification phenomena and related effects, there gradually emerged the concept of electric charge, regarded as something like a liquid which could occur in two varieties, positive and negative, and of which a body could possess a varying quantity. Eventually it was realized that the force between two charges is proportional to the product of the charges (being repulsive for like charges and attractive for unlike ones) and inversely proportional to the square of the distance between them (Coulomb's law). Simultaneously, the properties of electric currents were being investigated, and they also were found to interact, with a force which again varied as the inverse square of the distance, but depended in detail on the relative orientation of the current-carrying wires (the Ampère and Biot–Savart laws). It gradually became apparent that it was conceptually simpler to formulate the laws of electricity and magnetism (as well as the law of gravity) in terms of *fields*, rather than direct interactions. For example, an electric charge was regarded as producing an 'electric field' whose strength was proportional to the inverse square of the distance from the charge; this field then exerted a force on any other charged body which happened to be around. Similarly, an electric current would produce a 'magnetic field' which would act on other currents (it was, of course, recognized that magnetic substances such as iron could also produce and be acted on by such fields, hence the name); and a massive body would produce a 'gravitational field'. Originally, these fields were visualized as some kind of physical distortion or disturbance of space; but, as we shall see, they gradually came to be thought of in more and more abstract terms, finally ending up (with the advent of quantum mechanics) as little more than the potentiality for something to happen at the point in question. It was appreciated early on, of course, that the current whose magnetic effects could be detected was nothing but the flow of electric charges, and later, that it was proportional to the electric field in the conductor carrying it (Ohm's law). A second crucial link between electricity and magnetism came with Faraday's discovery that the motion of a magnet could induce currents in a conductor, or, in terms of fields, that a varying magnetic field automatically produced an electric field.

In optics, a long debate took place between those who believed that light was a stream of particles (the view favoured by Newton) and the school which held it to be a form of wave motion analogous to waves on water. The proponents of the wave theory could point to the phenomenon of interference as evidence; and since this phenomenon is so fundamental to modern atomic and subatomic physics, it is as well to take a moment to explain it, using a rather artificial, but very well-worn, example. Suppose we have a source of particles—for example, bullets—positioned some distance behind a screen S_1 in which two slits are cut (Figure 1.1). A second screen (S_2) provides a way of catching and registering the particles. We may assume that the particles bounce off the walls of the slits in some fairly random way which may

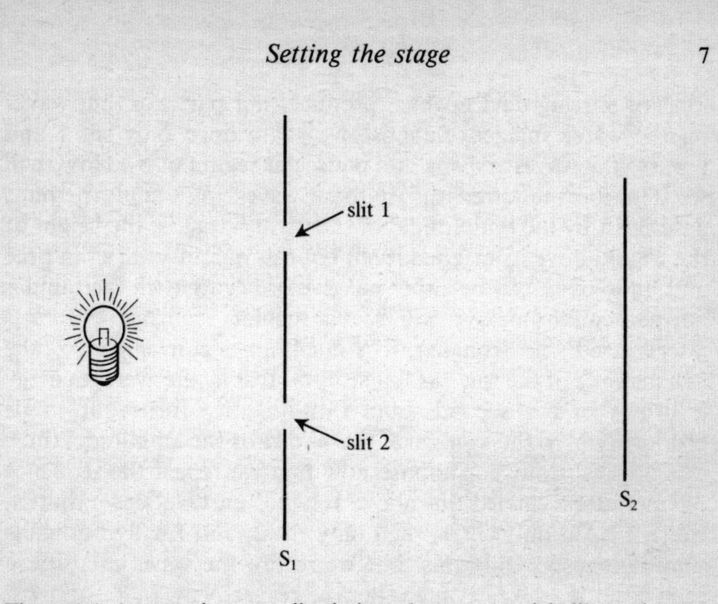

Figure 1.1 An experiment to discriminate between particle-like and wave-like behaviour

not be calculable or predictable in detail. However, we can make one very firm prediction which is quite independent of such details. Suppose we count the number of particles arriving per second over some small region of S_2, first with only slit 1 open, next with only slit 2 open and finally with both slits open. The source is supposed to be spewing out particles at the same rate throughout. Moreover, we assume (this can be checked if necessary) that the act of opening or closing one or other of the slits does not physically affect the other slit, and that there are not enough particles in the apparatus at the same time to affect one another. Then, if N_1 is the number arriving when only slit 1 is open, N_2 the number when only slit 2 is open, and N_{12} the number when both are open, it is immediately obvious that we must have $N_{12}=N_1+N_2$. That is, the total number of particles arriving at the small region of S_2 in question must be equal to the sum of the numbers of those which would have arrived through each slit separately.

Now suppose we have the same set-up as regards the screens and the position of the source, but that they are now partially immersed in a bath of water, and that the source is in fact some kind of stirring device which produces not particles, but waves on the water surface. Suppose we again open only slit 1 and observe the waves arriving at a particular region of S_2. How shall we describe the 'strength' of these waves? We could perhaps describe it by the 'amplitude' of the wave—that is, the height of the water above its original level; but this quantity varies in time and, moreover, can be either positive or negative (in fact, under normal conditions we will see a regular sequence of crests interspersed with troughs). It is much more convenient to take as a measure of the wave its 'intensity'—that is, the average energy it brings in per second—and this quantity turns out to be proportional to the average of the *square* of the amplitude; thus, it is always positive. Suppose now that we repeat the sequence of operations carried out above for the 'particle' case: that is, we first open only slit 1, then only slit 2, and finally both slits simultaneously. Only this time we record the wave intensity in each case: let the relevant values be respectively I_1, I_2, and I_{12}. Do we then find the relationship $I_{12} = I_1 + I_2$? That is, does the intensity of the wave arriving when both slits are open equal the sum of the intensities arriving through each slit separately? In general, it certainly does not. In fact, there will be some points on the screen where, although the intensity is positive with either slit alone open, it is zero (or nearly so) when both are open! In the case of a water wave it is very easy to see why this is so, for it is the *amplitude* (height) of the wave which is additive—that is, when both slits are open, the height of the wave at any given point at a given time is the sum of the heights of the waves arriving through each slit separately. Since the amplitudes add, the intensities, which are proportional to the squares of the amplitudes, cannot simply add. In fact, if A_1, A_2, and A_{12} denote the amplitudes under the various conditions just mentioned, we have $A_{12} = A_1 + A_2$, and therefore

$$I_{12} = A_{12}{}^2 = (A_1 + A_2)^2 = A_1{}^2 + A_{12}{}^2 + 2A_1A_2$$

whereas the sum of I_1 and I_2 is just $A_1{}^2 + A_2{}^2$. So the two expressions differ by the term $2A_1A_2$. In particular, if A_1 happens to be equal to $-A_2$—that is, if the wave arriving at the point in question through slit 1 has a crest just where that coming through slit 2 has a trough, and vice versa—then I_1 and I_2 are each separately positive, but I_{12} is zero. We say that we are seeing the effects of 'interference' (in this case 'destructive' interference) between the waves arriving through the two slits. The phenomenon of interference is characteristic of wave phenomena, and can be readily observed with all sorts of waves: water waves, sound waves, radio waves, and so on. However, it is important to observe that the essential characteristic of a 'wave' in this connection is not that it is a regular periodic disturbance, but that it is characterized by an amplitude *which can be either positive or negative*; without this feature there would be no possibility of destructive interference, at least. The actual physical nature of the amplitude depends on the wave, of course. In the case of water waves it is the height above background; for sound waves it is the deviation of the air density or pressure from its equilibrium value; for radio waves it is the value of the electric field; and so on. But in each case there is some quantity which can have either sign; and in each case, since we wish the intensity of the wave to be a positive quantity, it is taken to be proportional to the *square* (or average of the square) of this quantity. Actually, in each case it is possible to define the intensity simply as the flow of energy; then, since the overall energy of wave motion is conserved (see below), it follows that if the interference is destructive in one region, there must be other regions in which it is 'constructive'—that is, in which I_{12} is greater than the sum of I_1 and I_2.

The first half of the nineteenth century saw more and more experiments in optics which could apparently only be easily explained in terms of interference, and hence pointed to the interpretation of light as a wave phenomenon. Nevertheless, perhaps in part because of the towering prestige of Newton, there remained a substantial number of eminent scientists who held that light was a stream of particles. In the history of their gradual conversion, one episode is particularly amusing. In an attempt to refute a theory of diffraction developed by the wave theorists, it was pointed out by Poisson, one of the advocates of the particle picture, that this theory led inevitably to the prediction that in the centre of the shadow of an opaque circular object there should appear a bright spot—which, he claimed, was patently absurd. The proponents of the wave theory promptly went away and did the experiment, and showed that such a spot did indeed exist! Such instances of being hoisted with one's own petard are not uncommon in the history of physics. By around the middle of the nineteenth century few people doubted that light was indeed a wave—a conclusion which, as we shall see later, was somewhat premature.

A third major area of development in this period was thermodynamics—that is, the study of the relationships among macroscopically observable properties of bodies such as pressure, volume, temperature, surface tension, magnetization, electric polarization, and so on. One major theme which ran through this development was the concept of conservation of energy. In mechanics the 'kinetic energy' of a body—that is, the energy resulting from its motion—is conventionally defined as one-half its mass times the square of its velocity, and its 'potential energy' as the energy which it has as a result of the forces which can act on it; for example, a body moving in the uniform gravitational field of the earth near its surface has a potential energy equal to its mass times its height above the ground times the constant of gravitational acceleration (roughly 10 metres per second per second). For motion in a more general gravitational field (for example, for the planets moving in the field of the sun, which is not uniform) or when the forces are not gravitational in origin, one can often make a natural extension of this definition of potential energy. Its main property is that it depends only on the position of the body, not on its velocity or acceleration. The point of defining kinetic and potential energy in this way is that for an isolated system acted on only by forces such as gravitation and electricity, it is a consequence of Newton's laws that although the two forms of energy can be converted into one another, their *sum*—that is, the total energy—remains constant. This is known as the 'law of conservation of energy'. Physicists love conservation laws, both because of their intrinsic elegance and simplicity and because they often make calculations much simpler. Indeed, Newton's original first law simply states the law of conservation of momentum (mass times velocity) for a body subject to no forces; while his third law, when combined with his second, states in effect that for a system of bodies interacting only with one another the *total* momentum is conserved. We shall meet many other examples of conservation laws below.

It was natural, therefore, to try to extend the law of conservation of energy not only to wave phenomena (for which it works well) but also to the realm of thermodynamics. In some cases this was obviously quite plausible. For example, a mass hung on the end of a light spring and allowed to oscillate will satisfy the principle of energy conservation, at least for a short time, provided that we extend the definition of potential energy to include a term associated with the stretching of the spring. However, there are at least two obvious problems. In the first place, if the mass is subject to any frictional forces at all—for example, from the viscosity of the surrounding air—it will eventually come to rest, and it is then easy to check that its total energy (the sum of kinetic and potential terms) has decreased in the process. Second, there are clearly cases in which a system appears to *gain* energy for no obvious mechanical reason: for example, if I fill a coffee tin with water, jam the lid on hard, and heat the tin over a flame, then the lid will eventually be blown off with considerable velocity—that is, with a considerable gain in kinetic energy, despite the fact that there is no loss of potential energy associated with the process. With the help of examples such as these it was eventually realized that heat—originally thought of as a fluid, 'caloric', somewhat similar to the contemporary concept of electric charge—was actually nothing but a form of energy—the kinetic and potential energy of the random motion of the molecules which by that time were becoming accepted, mainly on the basis of chemical evidence, as the microscopic building blocks of matter. Once it became possible to measure heat quantitatively and the conversion factor between the units of heat as conventionally measured and those of mechanical energy was found, it became clear that the total energy—that is, heat plus mechanical energy—is indeed conserved; thus, for example, in the case of the oscillating mass on the spring, the mechanical energy 'lost' in friction is in fact merely converted into an equivalent amount of heat. This generalized version of the law of conservation of energy is known as the first law of thermodynamics. The better-known second law expresses the fact that although heat may be converted into mechanical work (or other useful forms of energy) to a certain extent, there are strict limitations on the efficiency of this process; in particular, there is a certain function of the state of a system—its 'entropy'—such that the total entropy of the universe—by which in this context is meant the system plus whatever interacts with it—must increase or, at best, remain constant. Although a body does not contain a fixed 'amount of heat'—heat is only one form of energy, and can be exchanged with other forms—the heat *added* to a body is related to its change in entropy, so it is natural to think of entropy as a measure of the amount of disordered, random molecular motion. If we do so, then the second law of thermodynamics expresses the intuitively plausible fact that it is possible to convert 'ordered' motion—as in the oscillations of the mass on the spring—into 'disordered' motion—the random motion of the molecules—but not to reverse the process, at least on a global scale. Thus, if a body is cooled in a refrigerator, its entropy thereby being reduced, other parts of the universe must experience a compensating increase in entropy—in this example, the room is heated. As we will see, statistical mechanics later gave a more quantitative foundation to the interpretation of entropy as a measure of disorder.

The second half of the nineteenth century saw two major unifications of apparently disparate branches of physics, as well as crucial progress towards a third. To take the latter first, it had long been recognized that each of the various chemical elements and compounds, at least when in gaseous form, emitted light not at random but only with certain specific wavelengths characteristic of the element or compound in question. Since the different wavelengths are bent by different amounts by a glass prism, they appear on the photographic plate of a simple spectrometer as isolated vertical lines. The specific set of wavelengths emitted by a substance is known as its (emission) 'spectrum', and the experimental study of these patterns as 'spectroscopy'. Its practitioners rapidly learned to identify the chemical components of unknown substances by the set of spectral lines emitted, a technique which is nowadays quite essential, not only to chemical analysis but also to astrophysics (see Chapter 3). It soon became clear that there were a number of regularities in the spectra of the elements. In particular, in the case of hydrogen, which on the basis of chemical evidence was recognized as the simplest element, there were a number of apparently quite precise numerical relationships between the various wavelengths emitted, involving only integral numbers. Although the count of these numerical relationships grew rapidly, their origin remained quite obscure as long as no microscopic picture of the atom was available. In addition, the properties of so-called blackbody radiation—the radiation found when a cavity with completely absorbing ('black') walls is maintained at a given temperature—were studied in detail, with results which were puzzling in the extreme.

The first major unification which was actually achieved in the latter half of the nineteenth century was of optics with electromagnetism (and, in the process, of electricity and magnetism with one another). Newton's mechanics and gravitational theory, and subsequent theories of electric and magnetic effects, had implicitly relied on the notion of 'instantaneous action at a distance'—that is, that the gravitational force, for example, exerted by one astronomical body would be felt instantaneously by a second body distant from it, without any need for a finite delay in transmission. The question of how the force came to be transmitted instantaneously, or indeed at all, was regarded as meaningless: instantaneous action at a distance was simply a basic fact of life which did not itself require explanation. Certainly there were people who felt that this approach was unsatisfactory metaphysically, and who hankered after a detailed account of the way in which the force was propagated; but the exponents of action at a distance could point out, correctly, that while the concept might look a little odd and even contrary to 'common sense', calculations based on it always seemed to give good agreement with experiment. (We will see in a later chapter that a similar situation obtains today with regard to the theory of measurement in quantum mechanics.) At the same time, it had been known since the end of the seventeenth century, from astronomical observations, that the speed of light was finite (and indeed the value inferred was quite close to that accepted today, approximately 3×10^8 metres per second); however, no special significance was attached to this fact, since there was no particular reason to believe that light played any special role in the scheme of physics.

In the 1860s, in the course of thinking about the laws of electricity and magnetism as then known, the British physicist J. C. Maxwell noticed an odd asymmetry: namely, that a changing magnetic field could induce an electric field, but, as far as was known, not vice versa. Maxwell argued from consistency that this could not be the case, and that there was a missing term in the equation describing the observed phenomenon, and once this was assumed a remarkable conclusion emerged: electric and magnetic fields could propagate through empty space as a kind of wave motion. Moreover, if one put into the equations the constants measured in laboratory experiments on electricity and magnetism, the velocity of such a wave could be calculated—and turned out to be just about the velocity of light! The obvious conclusion was that light is nothing but an electromagnetic wave—that is, a wave in which electric and magnetic fields oscillate in directions perpendicular to one another and to the direction of propagation of the light. The phenomenon of 'polarization', which had long been known experimentally, could then be easily understood: the plane of polarization is simply the plane of the electric field. We now know, of course, that visible light is just one small part of the 'electromagnetic spectrum', corresponding to waves with wavelengths in the range detectable by the human eye (approximately 4×10^{-7} to 8×10^{-7} metres); as far as is known, electromagnetic waves can propagate with any wavelengths, λ, and the associated frequency, $\nu = c/\lambda$ (where c is the speed of light), from waves in the kilohertz (kHz) frequency region emitted by radio stations to the 'hard γ-rays' with frequencies of the order of 10^{28} hertz (Hz) observed in cosmic radiation. The conventional division of the electromagnetic spectrum is shown in Figure 1.2.

One immediate consequence of the theory of electromagnetic fields as developed by Maxwell is that the velocity of propagation of electromagnetic effects is finite—equal, in fact, to the velocity

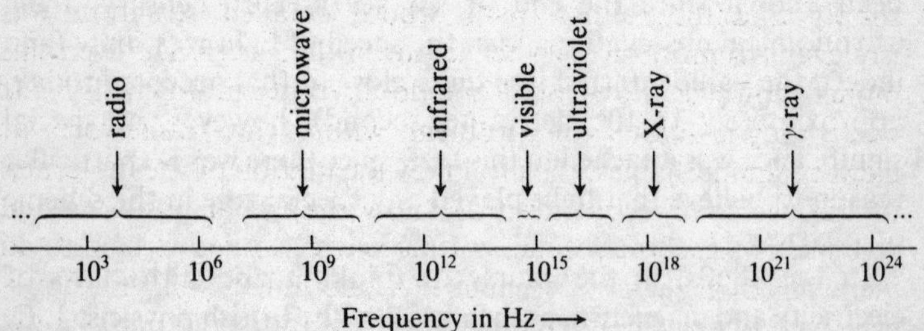

Figure 1.2 The conventional division of the electromagnetic spectrum (approximate; note that the scale is logarithmic, that is, each marked division corresponds to an increase in frequency by a factor of 1000)

of light, c. If, for example, we turn on a switch in the laboratory and thereby set off a current in a circuit, the magnetic field produced by that current at some point a distance R away is not immediately felt there, but takes a time R/c to materialize; it is only because the velocity of light is so enormous relative to everyday speeds that we have the illusion of instantaneous action at a distance. Thus, a question which within the framework of the original Newtonian scheme was 'merely philosophical'—How is action at a distance possible?—has now acquired not only a physical meaning but a very satisfying answer, at least as regards electromagnetism: namely, that it isn't possible; the fields propagate the interaction from each point to its immediate neighbourhood at a finite speed.

The second great unification of the late nineteenth century was of the molecular theory of matter with thermodynamics. It had long been accepted that simple chemical substances were built out of small identical building blocks, molecules, and that the molecules in turn were composed of sub-units, atoms, with each chemical element having its characteristic type of atom; and that chemical reactions could be interpreted in terms of the breakup of molecules and the rearrangement of atoms to form new ones. The precise nature of these atoms and molecules was quite obscure; however, there were some situations where, plausibly, their detailed structure might not matter too much. For example, it was known that in a reasonably dilute gas (such as air at room temperature and pressure) the molecules are quite far apart compared with their typical packing in a liquid or solid; so it seemed reasonable to think of them to a first approximation as simply like tiny billiard-balls whose detailed structure was irrelevant to most of their behaviour in the gas. Even given this model, however, how could one do anything useful with it? If one thought about the problem within the framework of Newtonian mechanics, then to know the motion of the molecules of the gas, one would need to know the initial positions and velocities of each molecule, and moreover, to know them to an incredibly high degree of accuracy. (As any billiards-player knows, even with two or three colliding objects a tiny change in the initial conditions can rapidly lead to a huge change in the subsequent motion!) The eighteenth-century French mathematician Laplace had indeed contemplated, as a sort of philosophical thought-experiment, a being who had the capability of acquiring all this information, not just for a gas but for the whole universe, and had concluded that to such a being both the past and the future of the universe would be totally determined and predictable. (This view, sometimes known as 'Laplacean determinism', has some curious echoes in modern times.) However, whether or not one accepts the outcome of the argument, in practice it is totally unrealistic to pose the problem in this way; we can neither acquire the necessary information nor process it—that is, solve the Newtonian equations of motion—in any reasonable time; and no advance, however spectacular, in the computing power available to us in the remotely foreseeable future is likely to change this state of affairs.

Into this situation there comes to the rescue the discipline we now call 'statistical mechanics'. It starts with the crucial recognition that not only is it impossible in practice to know the detailed behaviour of each individual molecule, but that in any case there would be no point in doing so. Most of the properties of macroscopic bodies which we can actually measure—magnetization, pressure, surface tension, and so on—are in fact the cumulative outcome of the action of huge numbers of molecules, and it is a general result of the science of statistics that such averages are very insensitive to details of individual behaviour. As an example, let us consider the pressure exerted by a gas such as steam on the walls of the vessel containing it. This pressure is nothing but the sum of the forces exerted by the individual molecules as they collide with, and are reflected by, the walls; the faster the incident molecules, the greater the force. Now any one molecule may come up to the wall slowly or fast, and therefore exert a small or a large force: but since the measured pressure arises from a vast number of such individual events, the fluctuations rapidly even out, and the resulting pressure is usually constant within the accuracy of our measurements. Thus we do not need to know the detailed behaviour of each molecule: all we need is *statistical* information—that is, information about the *probability* of a randomly chosen molecule having a given velocity. This is what it is the business of statistical mechanics to provide.

How does statistical mechanics set about providing this information? Historically there have been two main approaches, the 'ergodic' and the 'information-theoretic'; while they give the same answers, at least as regards the standard applications of statistical mechanics, the way in which the basic assumptions are formulated look rather different. Both approaches rest on the fact that there are very few simple macroscopic variables of the system which are conserved in the motion. For example, for an isolated cloud of gas the only such conserved quantities are the total energy, total momentum, and total angular momentum;[1] for a gas enclosed in a thermally insulating flask, only the total energy; and so on. There are indeed known to be other quantities which must be conserved, but they are generally very complicated, and the hope is usually that by ignoring them one will not get misleading results.

The ergodic approach proceeds by arguing, crudely speaking, that since the dynamics of a macroscopic system of particles is extremely complicated, it is plausible to say that whatever state it starts in, it will in time pass through all other states which are compatible with the conservation laws (for example, in the case of the gas in the flask, all states of the same total energy as the initial one). Since in a normal experiment on a macroscopic system we are in effect averaging the observed quantities over times which on an atomic scale are very long, it is argued that we are in effect averaging over all accessible states of the system. The main problem with the ergodic approach is that while the basic hypothesis—that the system will in time pass through all accessible states—may be thought to be intuitively plausible, it has been actually proved only for an extremely small class of systems, none of which are particularly realistic from an experimental point of view.

The information-theoretic approach avoids this problem by simply postulating from the start that all states which are compatible with the knowledge we have about the system are equally probable. For example, suppose we have a fairly small (but still macroscopic) subsystem which is in thermal contact—that is, can exchange heat energy—with a much larger system (the 'environment'), the whole being thermally insulated from the outside world. (An example which approximately fulfils these conditions might be a carton of milk in a refrigerator.) Then we know that the energy, E, of the whole must be constant, but that that of the subsystems need not be, and we can ask the question: What is the probability, P_n, that the subsystem is in some particular state, n, which has energy E_n? The answer is in principle very simple: it is simply proportional to the number of states available to the environment which are compatible with this hypothesis—that is, which have energy $E - E_n$. One might think that it would be very difficult to determine what this number is, but it turns out that for a macroscopic system it is actually rather easy to get an approximate expression which is negligibly in error provided that the number of particles is sufficiently large. The final answer is that P_n is a simple function of the energy E_n—namely, a constant times exp $(-E_n/E_0)$, where E_0 is a quantity of the dimensions of energy which characterizes the state. (This answer—the so-called Boltzmann distribution—is also obtained rather more indirectly from the ergodic approach.) If one then compares the predictions which follow from this expression for macroscopically measurable quantities with the formulae of thermodynamics, it turns out that E_0 is nothing but the temperature, provided that the latter is measured in appropriate units with an appropriate origin. Nowadays in physics it is conventional to take the origin 'appropriately', that is, to measure temperatures on the so-called absolute, or Kelvin, scale, on which the absolute zero of temperature corresponds to about -273 degrees Celsius; however, it is conventional, though somewhat illogical, to continue to measure temperature in degrees rather than in units of energy, so one needs a constant to relate E_0 to the conventional temperature in degrees. In fact one has to write $E_0 = k_B T$, where the quantity k_B, known as Boltzmann's constant, is numerically about 1.4×10^{-23} joules per degree. It is important to appreciate that k_B, unlike the speed of light, c, and Planck's constant, h, which we will meet below, really has no fundamental significance, any more than does the constant we require to convert feet into metres; in fact, had the history of physics been different, we might well have done without separate units of temperature altogether, merely measuring 'hotness' in units of energy.

Armed with the Boltzmann distribution, and the techniques for counting states which led to it, we can not only make detailed correspondences between all the formulae of thermodynamics and the results which follow from statistical mechanics, but in many cases, if we are prepared to assume some specific microscopic model, we can actually calculate the quantities which in thermodynamics have to be taken from experiment. For example, if we take as a model of a gas a set of freely moving billiard-ball-like molecules, we can justify the classical phenomenological laws of Boyle and Charles relating pressure, volume, and temperature. This kind of demonstration of correspondence between the microscopic picture and the macroscopically measured quantities lies at the basis of the bulk of modern work in the physics of condensed matter (though I shall argue in Chapter 4 that it is easy to misinterpret its significance). One quite general result emerges from a comparison of the thermodynamic and statistical-mechanical formulae. The mysterious quantity entropy, which was originally introduced in a thermodynamic context in connection with the second law, turns out to be related to the number of states effectively available to the system subject to what we know about it. That is, it can be interpreted intuitively either as a measure of the 'disorder' of the system, or, more strikingly, as a measure of our ignorance about it. That a quantity which at first sight has such an anthropomorphic interpretation can actually play a role in the 'objective' thermodynamic behaviour of the system is rather striking and puzzling, and is related to some even more unsettling questions which I shall take up in Chapter 5.

With these major advances in hand, the physics community entered the twentieth century in a mood of high confidence. All the pieces of the jigsaw seemed to be coming together: Newtonian mechanics was a complete description of the motion of all possible massive bodies, from planets down to atoms; Maxwell's theory of electromagnetism not only explained all of optics but held out the prospect of understanding the interactions, presumed to be mainly electrical, between atoms and molecules; and statistical mechanics would allow the explanation of the properties of macroscopic bodies in terms of those of the atoms composing it. True, there were puzzles left: there was still no understanding of the detailed structure of atoms or of the mysterious regularities involved in their spectra; and while there was a good qualitative, and even quantitative, microscopic theory of the macroscopic behaviour of gases, no such understanding was yet available for liquids or solids. But these were matters of detail. Who could doubt that the framework of physics was sound, and that these puzzles would eventually find their explanation within it? The prevailing mood of the period was expressed by the British physicist Lord Kelvin, who, in a lecture in 1900 surveying the state of physics as it entered the twentieth century, concluded that all was well apart from two 'small clouds' on the horizon (of which more below); in all other respects there was no reason to doubt that the subject was firmly set in the right direction. It may be a good antidote, in 1987, to some of the more breathless popular accounts of recent advances in particle physics or cosmology to put oneself in the position of our forerunners in 1900 and to try to appreciate how very firm and unshakable the basis of their enterprise must have seemed to them, and how unthinkable the idea that their whole conceptual framework might be in error.

Physicists, alas, do not always make good prophets. In the event, within little more than two decades, Lord Kelvin's small clouds had each blown up into a major hurricane: the whole edifice of classical physics was in ruins, and the very questions which were being asked would in some cases have seemed nonsensical to nineteenth-century physicists. Indeed, if one excepts perhaps the years of Newton's great work, the first thirty years of the twentieth century were by any reasonable account the most exciting period, and certainly the most revolutionary, in the whole of the history of physics. One of Lord Kelvin's clouds was the experiment of Michelson and Morley, which had failed to show the expected dependence of the velocity of light on the motion of the observer; this led—logically, even if not historically—to Einstein's special theory of relativity, and eventually his general theory. The second cloud concerned the specific heat of polyatomic molecules, and this was resolved only with the development of quantum mechanics. Both these theories—special and general relativity, and quantum mechanics—violently challenged the whole conceptual framework in which classical physics had been formulated.

I will discuss the special and general theories of relativity in Chapter 3 in the context of their cosmological applications, and here only summarize a few of the features of the special theory which are relevant to the topics to be mentioned in the next chapter. These are: that the velocity of light, c, is independent of the frame of reference in which it is measured, and is a fundamental constant of nature; that this velocity is the upper limit of velocity of propagation of any kind of physical effect, and that while bodies of zero mass (such as light itself) automatically travel at speed c, no body of finite mass can in fact be accelerated up to this velocity; that energy, E, and mass, m, are interconvertible according to Einstein's famous relation $E = mc^2$; that for isolated systems, the laws of conservation of energy and momentum are not two different principles, but the same principle viewed from different frames of reference, and that in any frame of reference, energy, E, and momentum, p, are related by the equation

$$E^2 = c^2p^2 + m^2c^4$$

(the famous $E = mc^2$ being the special case appropriate to a particle at rest—that is, with $p = 0$); and that 'moving clocks appear to run slow', that is, that a physical phenomenon such as the decay of a particle which occurs at a given rate when the particle is at rest appears to an observer with respect to whom the particle is moving to take place at a reduced rate.

To introduce quantum mechanics, let us digress for a moment to a third major advance of the first three decades of the twentieth century, which although not in itself as revolutionary from a conceptual standpoint as relativity or quantum theory nevertheless plays a central role in modern physics: the theory of atomic structure. A major clue had come in the last decade of the nineteenth century with the discovery that electric charge, or at least the negative kind of electric charge, was not a continuously divisible fluid, but came in discrete units, in fact in the form of the particles we now call 'electrons'. The electron charge, e (about 1.6×10^{-19} coulombs), is a fundamental constant of nature. While it was clear that electrons were a constituent of atoms, and therefore that the latter, being electrically neutral, had to contain also some positive charge, the nature and disposition of this positive charge remained obscure. It was clarified when Lord Rutherford, in a famous series of experiments, discovered that the positive charge was concentrated in a tiny blob—what we now call the 'nucleus'—in the centre of the atom, which contained all but a very small fraction of the atom's mass. Since the nuclear charge itself appeared to occur only in multiples of e, one was led to deduce the existence of a positively charged particle with charge e and mass about two thousand times that of the electron—what we now call the 'proton'. The general picture of the atom which emerged was of a central nucleus, at most a few times 10^{-15} metres in diameter, containing protons and (with the virtue of hindsight!) possibly other things, surrounded by a swarm of electrons which orbit the nucleus at a distance of about 10^{-10} metres. The electrons orbiting the small central nucleus behave very much like the planets orbiting the sun, the main difference (apart from scale) being that they are kept in orbit by electrostatic, rather than gravitational, attraction; in fact, the model of the atom which emerged from Rutherford's experiments is often called the 'planetary' model.

But planetary motion involves acceleration, and according to classical electrodynamics, accelerating charges radiate (the principle that is involved in radio transmission), which in the case of the atom leads to two major problems. First, one would expect the electrons to be able to radiate light of any frequency, whereas the spectroscopic evidence shows that any given atom radiates only at certain special frequencies. Second, by radiating, the electrons lose energy, which means that they should rapidly spiral down into the nucleus—that is, the atom should collapse. To solve these problems, the Danish physicist Niels Bohr introduced a hypothesis which, within the framework of classical mechanics, looked totally arbitrary and unjustifiable: namely, that the electrons were constrained to move only in certain particular orbits, and that when they jumped between these allowed orbits, the frequency, ν, of the emitted radiation was related to the difference in energy of the orbits, ΔE by the relation $\Delta E = h\nu$. In this formula h represents a constant which had been introduced by Planck a few years earlier in the theory of blackbody radiation, and is named after him ('Planck's constant'); it has the dimensions of energy times time and is numerically about 6.6×10^{-34} joule-seconds. It is essential to appreciate that h, like the speed of light, c, but unlike Boltzmann's constant, k_B, is a genuinely fundamental constant and not just a conversion factor between different arbitrarily chosen scales:[2] while we are, of course, free to choose our scales of length, time, and mass (and hence energy) so that h and/or c take the numerical value 1 (a practice which is common in modern atomic and particle physics), according to Bohr phenomena in mechanics which involve values of the product of energy and time comparable to h or smaller (such as the motion of electrons in atoms) appear qualitatively different from those where this combination is large compared to h (as in the motion of the planets of the solar system). The occurrence of such an 'intrinsic scale'—that is, the fact that, if we measure the combination $E \times t$ in units of h, the behaviour is quite different for $E \times t \gg 1$ and for $E \times t \ll 1$—is totally alien to classical Newtonian mechanics, and it is worth emphasizing (with an eye to some speculations I shall make later) that there is absolutely no way in which, simply by inspecting classical mechanics 'from the inside', one could ever have deduced the existence, or indeed the possibility, of such a scale. In the end it was forced on us by experiment, and can be understood only in the light of a totally new theory which rejects the whole classical framework—quantum mechanics, to which I now turn.

Let us proceed, as is common practice in introducing new concepts in physics, by considering an experiment which from a practical or historical point of view is rather artificial, but which encapsulates in a simple and clear way features which in real life are inferred from more indirect and complicated experiments—the 'two-slit' set-up described above and illustrated in Figure 1.1. We saw that if we fired ordinary particles such as billiard-balls through this apparatus, then no matter how complicated the scattering at each slit, the total number of particles arriving at a given point on the screen S_2 when both slits are open is just the sum of the numbers which arrive through each slit separately when it alone is open; if, on the other hand, we propagate a wave (such as a surface wave on water) through the apparatus and measure the intensity arriving at a given point on S_2, then in general the intensity detected with both slits open is *not* the sum of the intensities arriving through each slit separately, but shows the interference effects characteristic of wave phenomena. As we saw, historically it was observations of this type which led to the conclusion that light was a wave phenomenon, rather than a stream of particles. Now suppose we do the experiment with (say) a source such as a hot filament which can emit electrons; and suppose that we turn the source down very low (or insert a filter between the source and S_1) so that the probability of more than one electron being in the apparatus at a time is negligible, and moreover use for S_2 a type of screen which will detect the arrival of each electron separately (this is in practice not difficult with suitable amplifying devices). What will we see? Well, first, we will confirm that as regards their arrival at S_2 the electrons do indeed behave like particles—that is, each electron on arrival gives a signal at one specific point on the screen; the signal is not spread out diffusely over a large area. So it certainly looks as if 'something' has arrived at a definite point, and at this stage there seems no good reason to doubt that the electron is indeed a particle in just the same sense as are billiard-balls. But wait: as we fire more and more electrons through the apparatus, keeping count of where they arrive, and plot the distribution of their arrival points, we begin to see that although the point of arrival of each electron appears to be quite random, a pattern is eventually built up; definitely, more electrons arrive in some areas of the screen than in others, and gradually it becomes clear that we are seeing an interference pattern very similar to the pattern we would get for water waves (or for light). It might then occur to us to check whether the number of electrons arriving at a particular point when both slits are open is indeed the sum of those arriving through each slit separately when it alone is open. The answer is no: indeed, there are points at which electrons arrive when either slit alone is open, but none arrive when both slits are open simultaneously. In other words, the electrons seem to be experiencing an interference effect precisely similar to that associated with wave phenomena.

We might then ask: if an electron can show wave-like aspects, can light show particle-like aspects? Indeed it can. If we do the same experiment with light, then, if our detection apparatus is of the fairly crude type normally used in, for example, secondary school optics experiments, then all we will see is the average light intensity arriving on a given area of S_2 over a fairly long period, and this will show the interference effects typical of a wave. But it is perfectly possible to refine our detection apparatus so that it can register very small intensities, and if we do this, we can verify that the light does not in fact arrive continuously, but in discrete chunks, or packets. That is, we can think of light, at least for some purposes, as after all consisting of a stream of particles; these particles are called 'photons'. Thus both our traditional 'particles' such as electrons and our traditional 'waves' such as light actually show some aspects of both wave and particle behaviour. This feature of quantum mechanics is known as 'wave–particle duality'.

We can make the picture a good deal more concrete than this. For definiteness let us concentrate on electrons (the case of photons, or of other particles such as protons, can be discussed similarly). Let us tentatively associate with an electron in a given state a wave, for the moment of unknown nature; and let us assume that this wave undergoes interference effects just like any other. Then, just as with the water waves discussed earlier, the amplitudes of waves coming by different paths will add—that is, $A_{12} = A_1 + A_2$—and the intensities, which we again take to be proportional to the average square of the amplitudes, will in general not add—that is, $I_{12} \neq I_1 + I_2$. If we now interpret the intensity of the wave as giving a measure of the *probability of detecting an electron* at the point in question, we will have at least a qualitative explanation of the observed interference effects. To make it quantitative, we will need to calculate the actual behaviour of the wave, and this requires us to make some correspondence between the wave and particle aspects of the electron's behaviour. The correspondence which has, in fact, been found to be successful is this: if the electron, viewed as a particle, is in a state with definite momentum p, then the associated wave has a definite wavelength, λ, which is related to p by the equation $\lambda = h/p$, where h is Planck's constant. This fundamental equation is known as the 'de Broglie relation'. From the formulae of special relativity it then follows that the energy, E, of the 'particle' is related to the frequency, ν, of the associated 'wave' by the relation $E = h\nu$ (and the same relation holds between the frequency of a classical light wave and the energy of the associated photons).

With the help of these relations, and the crucial assumptions made above about the interpretation of the wave intensity as the probability of finding a particle at the point in question, we can now explain *quantitatively* the observed interference pattern. Moreover, we can begin to understand the underlying reasons for the success of Bohr's postulates about the behaviour of electrons in atoms.

Consider, for example, an electron moving around the atomic nucleus in a circular orbit of radius r. Plausibly, the associated wave must fit back on to itself as we go once around the orbit—that is, there must be an integral number, n, of wavelengths, λ, in the circumference, 2πr: thus $n\lambda = 2\pi r$. But, by the above correspondence, this means that $pr = nh/2\pi$, and this was precisely the condition originally postulated by Bohr for an orbit to be 'allowed'. Moreover, when the electron jumps between levels, it generally emits a single photon; by energy conservation, the photon energy, E, must be equal to the difference between the energies of the electron orbits involved, ΔE, and so from the relation $E = h\nu$ above, we have $\Delta E = h\nu$ as postulated by Bohr. A quantitative account of the behaviour of the amplitude of the quantum-mechanical wave is given by the fundamental equation of non-relativistic quantum mechanics, Schrödinger's equation; but it should be emphasized that this amplitude has, itself, no direct physical interpretation—it is only the *intensity* which has a meaning, as a probability.

Viewed from the perspective of classical physics, quantum mechanics has many bizarre features. In the first place, it enables one to predict only the *probability* that an electron or photon will be detected at a particular point on the screen; just why a *particular* electron arrived at the point it did, and not somewhere else, is a question the theory cannot answer, even in principle. (This feature alone would have horrified most nineteenth-century physicists!) Second, it does not allow us to consistently attribute to a microscopic entity such as an electron a full and completely defined set of 'particle' properties simultaneously. For example, in classical physics we would think of a particle as possessing at any given time both a definite position, x, and a definite momentum p. But in quantum mechanics the probability of finding the particle is associated with a wave intensity, and a wave which has a perfectly well-defined wavelength is quite unbounded in spatial extent. If we wish the intensity to extend over only a finite region of space, we must make up a combination of waves of different wavelengths, and by the relation $\lambda = h/p$, there will then be a spread, or indeterminacy, in the value of the momentum, p, of the particle. This feature is summarized in the famous Heisenberg indeterminacy principle[3]

$$\Delta p \cdot \Delta x \geqslant h/4\pi$$

where Δp is an appropriately defined indeterminacy in momentum and Δx a similar indeterminacy in position. A similar relation, rather more subtle in interpretation, exists between the indeterminacy, Δt, in time t at which a process occurs (or, to put it rather crudely, how long it lasts) and the indeterminacy in the energy associated with it, ΔE: $\Delta E \cdot \Delta t \geqslant h/4\pi$. Thus, if an intermediate state of the system 'lasts' for a time Δt, its energy is indeterminate to within an amount of the order $h/4\pi\Delta t$, so that we may 'borrow' an energy of this order for it without violating the principle of energy conservation; an application of this principle is made in the next chapter.

Third, and most alarming of all, quantum mechanics requires, at least prima facie, that in the two-slit experiment the electron should behave in a qualitatively different way when it is left to itself—that is, when passing through the two-slit apparatus—and when it is observed (at the fixed screen). In the first situation it behaves like a wave, as it must to give the interference effects; while in the second it appears as a particle (that is, gives a single signal on the screen; a wave would give a spread-out, diffuse signal). The reader might well ask whether we can't arrange to observe it *during* its passage through the apparatus—for example, by placing a detecting device at each of the slits to see which one it actually went through? Indeed we can, and if we do, we always find that each individual electron did indeed come through one slit or the other, not both—but, then there is no interference effect in the intensity pattern. So in some curious sense the electron seems to behave as a wave as long as we don't observe it, but as a particle when we do! This is actually a special case of a much more general and, to many people, worrying paradox in the foundations of quantum mechanics to which I shall return in Chapter 5.

Over the last sixty years, the formalism of quantum mechanics, augmented by the generalizations necessary to accommodate special relativity and field theory, has had a success which it is almost impossible to exaggerate. It is the basis of just about everything we claim to understand in atomic and subatomic physics, most things in condensed-matter physics, and to an increasing extent much of cosmology. For the majority of practising physicists today it is *the* correct description of nature, and they find it difficult to conceive that any current or future problem of physics will be resolved in other than quantum-mechanical terms. Yet despite all the successes, there is a persistent and, to their colleagues, sometimes irritating minority who feel that as a complete theory of the universe quantum mechanics has feet of clay, indeed 'carries within it the seeds of its own destruction'. Just why this should be so will emerge in Chapter 5.

To conclude this introductory chapter, let me say a few words about the relationship between physics and other disciplines, and about the basic 'guide-lines' which most physicists accept in their work. As regards the distinctions between physics and its sister subjects such as chemistry or astronomy, these are mainly of historical significance. The subjects share, to a large extent, a common methodology and rely on the same basic laws of nature, and the fact that stars and complicated organic molecules tend to be studied in different university departments from crystalline solids or subatomic particles is an accident of the development of science. (Indeed, the fact that there is no firm dividing line is recognized in the formal existence of cross-border disciplines such as chemical physics and astrophysics.)

With respect to disciplines of a rather different nature, such as mathematics and philosophy, the relationship is more interesting. Historically, the development of physics and that of mathematics have been closely intertwined: problems arising from physics have stimulated important developments in mathematics (an excellent example is the development of the calculus, which was stimulated by the need to solve Newton's equations of motion); conversely, the prior existence of branches of apparently abstract mathematics has been essential to the formulation of new ways of looking at the physical world (for example, in the development of quantum mechanics). Most physicists today, particularly those whose business is theory rather than experiment, move backwards and forwards across the dividing line between the two disciplines so frequently that they are barely conscious of its existence.

Nevertheless, the distinction is an important one. Suppose a physicist is faced with a new type of phenomenon which he does not understand; for example, he (or a colleague) has measured the total magnetization of a piece of iron as a function of its 'history'—that is, of the operations carried out on it in the past—and has found it to behave in an unexpected way. As a very crude first approximation, we can separate his attack on the problem into two stages. At the first stage, he tries to identify what he thinks are the crucial variables and to formulate what he thinks are the correct relationships between them—that is, to build a plausible 'model' of the system. For example, he might decide that the crucial variables are the density of magnetization at each point of the sample and the local magnetic field and temperature at that point, and that these variables are related by a particular set of differential equations. In deciding that these are indeed the correct variables and the correct equations, he is in effect making an intelligent guess (see Chapter 4), and at this stage he is genuinely doing physics. However, it is very rare that the answer to the question he really wants to ask—how does the magnetization depend on history?—springs directly out of the equations he has written down, so he must now proceed to the second stage and go through the often complicated and tedious business of solving the equations to produce the required information. At this stage he is not doing physics but mathematics; in fact, in principle he could at this stage hand his equations over to a pure mathematician who had no inkling of what the symbols in them represent in the physical world, and the solution would be none the worse for that (though the physicist would, of course, have to tell the mathematician exactly what information he wanted to extract). Of course, this simple scheme is a considerable idealization of real-life practice; it is probably a reasonable approximation to what goes on in at least some areas of particle physics and cosmology, but is a much worse description of the typical situation in condensed-matter physics, where it is quite common for the physicist to set up a model, solve the relevant equations, find that the solution predicts results which when interpreted physically are clearly ridiculous (for example, that the magnetization increases without limit), realize that he has left some essential physical effect out, go back to stage 1, construct a new model and repeat the cycle, sometimes several times. In fact, the real-life situation is often even more subtle than this, as we will see in Chapter 4.

Nevertheless, although this effortless (and sometimes unconscious) switching between what I have called stages 1 and 2 is very typical of modern research in physics, it is important to bear in mind that the whole nature of the exercise at the two stages is really quite different. At stage 2 one is trying to extract, by the rigorous process of deduction characteristic of mathematics, the information which is in a sense already implicit in one's original equations; if one makes a mistake at this stage, one is not making a mistake about the real world but a *logical* error. At stage 1, on the other hand, one is trying to 'interface' one's mathematical description with the real world: if one gets things wrong at this stage, one has simply made a bad guess about how nature actually behaves. It is particularly important to keep the distinction in mind in the context of the increasingly widespread use of computers at stage 2; despite the appearance in recent years of a subdiscipline known as 'computational physics', computers don't do physics—they do applied mathematics—and the output of a computer is as good, or as bad, as the model fed into it by its human programmer as a result of his stage 1 considerations.

The question of the relationship between the disciplines of physics and philosophy is an intriguing and controversial one; I will touch on it, implicitly, in Chapter 5 and make only a few remarks here. Most contemporary working physicists themselves tend to use the word 'philosophical' (not infrequently accompanied by the word 'merely'!) to refer to questions concerning the very basic framework in which physics is done, and the language in which the questions are posed. Such questions might include, for example, whether all explanation in physics must in the last analysis be of the form 'A happens now because B happened in the past'; whether a theory (such as quantum mechanics) which refuses in principle to give an account of the reasons for individual events can be adequate; whether the conscious observer should be allowed a special role; and so on. Because such questions, of their nature, are not susceptible to an experimental answer (or at least, not so long as the experiments are interpreted within the currently reigning conceptual framework), the word 'philosophical' has become, in the argot of many contemporary physicists, more or less synonymous with 'irrelevant to the actual practice of physics'. Whether or not this point of view is a valid one is a question which will be raised again, by implication, in Chapter 5. It is interesting that, among those philosophers who have paid close attention to the conceptual development of modern physics, views seem to be divided on this point: one school feels that the very success of (for example) quantum mechanics shows that any a priori, 'philosophical' objections to the formalism must automatically fail, and that philosophy has to accommodate to physics rather than vice versa;

while another school believes that sufficiently strong a priori conceptual objections must brand even a successful physical theory as at best a transient approximation to the truth.

Let us now turn very briefly to some features of the general conceptual framework which most physicists accept, often unconsciously, when they think about the world. Needless to say, much of this framework is in no way peculiar to physics, but is characteristic more generally of those activities we would class as 'scientific'; but it is perhaps easiest to analyze it within the context of physics, and it is probably no accident that workers in subjects such as sociology who wish to make their subjects 'scientific' tend to look on physics as the paradigm to be imitated. Despite the fact that all the assumptions listed below may seem so obvious as to be hardly worth stating, almost every one of them has been seriously questioned at some time or other in the history of human thought; and indeed what seems to us common sense has been very far from the intellectual orthodoxy at various periods of history.

Perhaps the most basic assumption which underlies all of physics, and indeed all of science in the sense in which the word is currently understood, is that the world is in principle susceptible to understanding by human beings; that if we fail to understand a given phenomenon, then the fault is in us, not in the world, and that some day someone cleverer than ourselves will show us how to do it. Of course, exactly what is meant by 'understanding' is itself a subtle question, and I will return to it by implication in later chapters. However, it seems clear that unless we had some such belief, there would be little point in even trying to carry out scientific research at all.

A rather more specific assumption is that the world exhibits some kind of regularity and constancy in space and time: that the 'laws of nature' will not arbitrarily change from day to day, or from place to place. For example, a theory which held that the ratio of the frequencies of oscillation of two atomic clocks was quite different a million years ago from what it is today would probably have difficulty gaining acceptance, unless it could give a general formula for the change of the ratio with time and, preferably, make predictions about the results of experiments in the future.

As this remark suggests, another ingredient which physicists tend to require in anything they are willing to regard as an 'explanation' is predictive power. That is, the theory given should not only explain relevant facts which are already known; it should also be able to predict the results of some experiments which have not yet been carried out. If one thinks about it from a purely logical point of view, this is an odd requirement: the *logical* relation of the theoretical results to the experimental ones cannot depend on the temporal order in which they were obtained. But physicists are human beings, and everyone knows how much easier it is psychologically—and how much less fruitful—to generate a complete explanation for a set of existing experiments than to predict the result of a future one. Indeed, one of the tell-tale signs of a paper of the type physicists usually refer to as 'crackpot'—though it is certainly not peculiar to them!—is that such papers, while apparently able to explain a lot of existing data, rarely if ever venture to forecast the results of any experiment which has yet to be carried out. Needless to say, there are cases where the demand for predictive power cannot reasonably be met: in much of cosmology and even astrophysics, the events under consideration are either by their very nature in the past or so far in the future as to be beyond experimental reach; and, as we have seen even in more mundane physics, the demand for *exact* predictability as regards individual events has had to be dropped with the advent of quantum mechanics. Still, there are few things which make the average physicist sit up and take notice more than a clean prediction of a qualitatively new phenomenon—for example, an elementary particle of a new type.

It is almost too obvious to be worth stating that physics is an experimental subject. Yet, it is worth remembering, perhaps, that it would not have seemed at all obvious to our medieval ancestors, most of whom would have regarded experimentation as essentially irrelevant to any of the important questions. (An attitude, one might add, which is not unknown even today among some denizens of departments of theoretical or 'mathematical' physics, where one can sometimes hear the word 'experiment' used without conscious irony to describe a computational solution of a purely mathematical problem! Compare the discussion above on the respective roles of physics and mathematics.) Indeed, it is often claimed that the rise of the 'experimental method' is synonymous with the rise of modern science. What kinds of assumptions do we, consciously or unconsciously, make about the nature and role of experiment?

In the first place, we assume that the results of experiments are *intersubjective*—that is, that they do not depend on who does the experiment or looks at the result. If you see a clock reading (say) 3.47 p.m., then I should be able to confirm that it does indeed read that time, and so should any other person who is familiar with the usual symbols and conventions. Of course, for all sorts of trivial reasons there are always going to be minor disagreements, but we implicitly accept that by using better spectacles, noise isolation, and so on, they can be reduced to as low a level as we like. It is significant that even the more bizarre interpretations of quantum mechanics, some of which I discuss in Chapter 5, feel it necessary to maintain this feature of experimental intersubjectivity. Indeed, in cases where agreement between different observers does not seem to be possible, scientists tend almost automatically to dismiss the phenomena in question as illusion and purely a product of the observer's mental condition. That is, we tend to divide the data of our immediate subjective experience into two classes: those which are reproducible by all competent observers and are the appropriate subject-matter of the so-called hard sciences such as physics, and those which are a function entirely of the observer's psychological state and have no 'objective' existence. It is amusing to reflect that the hypothesis that all experience falls into one class or the other is by no means logically necessary: in fact, one might whimsically call it the 'no-ghost' hypothesis. One might perhaps define a ghost, or ghostly apparition, as a phenomenon whose observation requires *both* the presence of certain objectively verifiable physical conditions (for example, an appropriate haunted house, the right anniversary, and so forth) *and* a certain mental state or mental qualities in the observer. (As is well known, many people, including the present author, appear totally incapable of seeing ghosts, whereas other equally reasonable people seem to have no difficulty in doing so—and this is not meant ironically.) Our present methodology of science implicitly assumes that ghosts, thus defined, do not exist; or more modestly, that if they do, they are not an appropriate matter for scientific investigation. It is not clear that our descendants in the twenty-fifth century will necessarily share this view.

A second assumption is that, in physics, we can only deal profitably with phenomena which are subject to some kind of quantitative measurement. In fact, many of the major advances in the history of physics have been associated with the invention of ways to measure (and hence to define quantitatively) concepts which previously were at best qualitative; as was mentioned earlier, Galileo's mechanics would have been impossible in the absence of relatively accurate clocks (invented in the comparatively recent past); and the development of quantitatively reliable thermometers was essential to the birth of the modern science of thermodynamics. It is quite conceivable that a major breakthrough in the study of highly organized—for example, biophysical—systems will require us to invent something like a quantitative physical (not mathematical!) measure of the (now) qualitative feature we refer to as 'complexity'.

It is natural that, once we limit ourselves to quantitatively measurable phenomena, we should also expect that it should be possible to describe them in the language of formal mathematics (there would, indeed, be little point in trying to devise quantitative measures if we could not take advantage of them). Indeed, for centuries the situation could be described simply: every one of the quantities we measured experimentally (distance, time, mass, and so on) was represented by a corresponding quantity in the mathematical model; and conversely, every symbol in the mathematics had a more or less direct interpretation in terms of experimental quantities. Already in the late nineteenth century, however, this simple picture began to be clouded: at least one important thermodynamic concept—entropy—turned out to have at best only a very indirect experimental interpretation, and the free-space electromagnetic field postulated by Maxwell, originally thought of in terms of mechanical vibrations of an all-pervasive ether, eventually turned out to be an elusive quantity which could at best be defined in terms of what *would* happen to a charged particle *if* there happened to be one there. Modern quantum mechanics has clouded the picture still further: according to most interpretations, the 'wave function', which is such a key element in the theoretical description, cannot be put into direct correspondence with anything at all in the observable physical world; we will explore in Chapter 5 some of the paradoxes and controversies associated with this feature.

After this all too sketchy preliminary survey, we will, in the next three chapters, examine the current state of affairs in some of the major areas of contemporary physics: high-energy physics, cosmology, and the physics of condensed matter.

2
What are things made of?

To the twentieth-century scientist and the twentieth-century layman alike, few things seem more natural or self-evident than the idea that the way to understand the properties and behaviour of a complex object is to take it apart into its constituent elements; and that if and when one reaches the level of a set of components which cannot be further subdivided, and understands the relationships among these components and the ways in which they interact, one will then have 'in principle' a complete understanding of the object as a whole. Perhaps this way of thinking has limitations, and I will return to this point in the conclusion; but it has certainly served us spectacularly well in the past, and it is no surprise that much of the glamour, as well as the cost, of modern physics continues to be associated with the search for the 'fundamental constituents' of matter—the enterprise known as particle physics or, for reasons which become clear, more commonly nowadays as high-energy physics.

As we all know, innumerable pieces of evidence from physics and chemistry lead to the conclusion that the ordinary matter we see all around us is composed of atoms,[1] and that each chemical element corresponds, crudely speaking, to a different type of atom. Each type of atom, in turn, is conceived as composed of a very tiny positively charged nucleus (with a radius of about 10^{-15} metres), which contains the bulk of the atom's mass, surrounded by a 'cloud' of negatively charged electrons which extends over a radius of about 1 ångström (10^{-10} metres). In the original, classical atomic model of Rutherford, as we saw, the electrons were conceived as orbiting the nucleus rather as the planets orbit the sun; but from the point of view of modern quantum mechanics, there really is no simple pictorial model which we can use, and all we can say is that an electron is liable to be found at random anywhere within a distance of about 1 ångström from the nucleus. The electrons are bound to the nucleus by the familiar Coulomb (electrostatic) attraction between unlike charges; however, they are readily exchanged or shared between atoms in chemical reactions (as, for example, when a chlorine (Cl) atom 'borrows' a spare electron from a sodium (Na) atom to form a molecule of common salt (Na^+Cl^-). It is conventional to measure the mass of an elementary particle in terms of the 'rest' energy to which, by Einstein's relation $E=mc^2$, it is equivalent. One MeV (million electron-volts) is the energy acquired by an electron when accelerated through a potential difference of a million volts; it is equal to 1.6×10^{-13} joules. In these units the electron has a mass (rest energy) of 0.51 MeV; it also has a (negative) charge of 1.6×10^{-19} coulombs (conventionally denoted $-e$), and an intrinsic angular momentum (spin) of $\frac{1}{2}$ in the natural units of $h/2\pi$. It is believed to be absolutely stable: that is, if left to itself in free space, an electron will never disintegrate.

The atomic nucleus is believed to consist of protons and neutrons, which collectively are called 'nucleons'. Both these particles are very heavy compared to the electron. The mass of the proton is approximately 938.3 MeV, and that of the neutron 939.6 MeV. Both have a spin of $\frac{1}{2}$ in natural units. The proton has a charge of $+e$—that is, as far as is known, its charge is exactly equal and opposite to that of the electron; the neutron, as its name implies, is electrically neutral—that is, it has no charge. Thus, the number of protons in the nucleus determines its total charge, and hence the number of electrons required to make the atom as a whole neutral; and this in turn determines the chemical behaviour; all atoms with the same number of protons in the nucleus therefore correspond to the same chemical element. The number of neutrons in the nucleus is more or less irrelevant to the chemical behaviour and affects mainly the mass of the atom. Atoms with the same number of protons but different numbers of neutrons are called different 'isotopes' of the same element. Thus, for example, the very simplest atom of all, ordinary (so-called light) hydrogen, has a nucleus which consists of a single proton. If we add a neutron, it becomes 'heavy hydrogen' or deuterium, which has (almost) the same chemical properties as ordinary hydrogen but is twice as heavy; if we then add a second proton, however, we get the light isotope of helium (^3He), which has totally different chemical properties.

The proton is believed (or more accurately, as we shall see, was until quite recently believed) to be absolutely stable. The neutron is not: if left to itself in free space, a neutron will eventually disintegrate, producing in the process a proton, an electron, and a third particle, which is uncharged and very difficult to detect directly, known as a neutrino (more accurately, an electron antineutrino). This disintegration is the prototype of what is known as a 'radioactive decay process': while for any one neutron the process of disintegration appears to occur in a totally random and unpredictable way, for a large assembly of neutrons there is a well-defined statistical behaviour, corresponding to a probability of decay of approximately $\frac{1}{15}$ per minute: that is, after one minute about $\frac{1}{15}$ of the neutrons in the sample will have decayed; after the second minute, about $\frac{1}{15}$ of the remainder, and so on. We say that the neutron has, in free space, a 'lifetime' of 15 minutes.[2] When the neutron is inside an atomic nucleus, the situation is more complicated. In some so-called radioactive nuclei it can decay much as in free space, though with a lifetime that can be anywhere from less than a millionth of a second to millions of years, depending on the nucleus; while in others it appears to be completely stable. To complicate things yet further, in other types of radioactive nucleus a *proton* can decay, producing, for example, a neutron and two lighter particles (an electron neutrino and a positron—see below).

The explanation for this apparently bewildering variety of behaviour lies in the fact that any process involving transformation or decay of particles into one another must satisfy certain *conservation laws*: that is, there are certain quantities which must have exactly the same values before and after the decay process. One very fundamental such law which we have already met in Chapter 1 is that of the conservation of energy, which is believed to apply to individual processes involving a few 'elementary' particles just as rigorously as it does to macroscopic bodies.[3] The total energy of a particle in free space is its rest energy, mc^2, plus the kinetic energy associated with its motion; since the latter is zero when the particle is at rest and positive otherwise, and since we can always choose to work in a frame of reference (see Chapter 3) in which the decaying particle is at rest, it immediately follows that a decay process can take place in free space only if the total mass (hence rest energy) of the 'decay products' (the particles produced in the decay) is less than that of the original particle. In the case of the decay of a neutron (mass 939.6 MeV) this condition is satisfied: the proton has a mass of 938.3 MeV, the electron 0.5 MeV, and the neutrino has zero rest mass,[4] so the total mass of the decay products is less than that of the neutron by 0.8 MeV, which is used up in kinetic energy. On the other hand, a proton clearly cannot decay in free space into a neutron plus anything else. In a nucleus the situation is different, however, because the total energy on each side of the equation must contain, in addition to rest energy and kinetic energy, the binding energy of the particle in the nucleus, and this can be different for a proton and a neutron in the same nucleus. Thus, depending on the quantitative value of the difference, we can get three different situations: a neutron is unstable, a proton is unstable, or both are stable.[5] Once the first neutron (say) has decayed, we of course have a different nucleus and must redo the calculation; eventually we will reach, perhaps after a series of decays, a nucleus in which both proton and neutron are stable and which is therefore not radioactive.

The above argument is, of course, not the whole story concerning the stability of the proton and the neutron. For example, why can the proton (mass 938.3 MeV) not decay into two electrons (total mass 1.02 MeV)? This decay is certainly not forbidden by energy conservation, but it *is* forbidden, *inter alia*, by another conservation law, which is believed to be equally universal—namely, the conservation of total electric charge (the proton has charge $+e$, the two electrons $-2e$). I will return to the general question of conservation laws later.

As we saw above, the force which holds the electrons in the atom is the familiar electrostatic attraction between unlike charges (the negatively charged electrons and the positively charged nucleus). With regard to the nucleons themselves, however, this force should have no effect on the neutrons, since they are uncharged, and should actually push the protons apart, since like charges repel. So, one might ask, how come that the nucleus stays together? The answer lies in a quite different force, the so-called strong interaction, which acts between nucleons, be they protons or neutrons, but has no effect on electrons, and which (at this level) is always an attractive force. As its name implies, the strong interaction is much stronger than the electromagnetic one (and can therefore overcome its effects), but has only a very short 'range' (about 10^{-15} metres, roughly the radius of a small nucleus); beyond this distance it falls off exponentially rather than, like the electromagnetic interaction, as the inverse square of the distance.

The three particles we have looked at so far—the electron, proton, and neutron—have in common, among other things, that they have finite rest masses (and therefore travel with a speed less than that of light), that they are stable for at least a reasonable time in free space, and that they can be deflected by electric and/or magnetic fields. (The neutron, which is uncharged, nevertheless has a finite magnetic moment and so can be deflected, albeit weakly, by a magnetic field.) In fact, while a complete picture necessarily involves quantum-mechanical considerations, in many situations each of these particles can be visualized as behaving somewhat like a tiny, hard billiard-ball—the picture which probably springs to most people's minds when they hear the word 'particle'.

The fourth 'particle' which plays an important role in everyday life, the photon, is rather different. We recall from Chapter 1 that when a light wave of frequency ν interacts with charged particles—for example, in the ionization process which is the first step in the blackening of a photographic plate—then the energy appears to be transferred in chunks, or 'quanta', of magnitude $h\nu$, where h is Planck's constant; and that it is these quanta which we call photons. Application of the (classical) relation between the frequency, ν, and wavelength, λ, of a wave, $\nu = c/\lambda$, and the (quantum) de Broglie postulate, $p = h/\lambda$, then indicates that the photon carries a momentum, p, related to its energy, E, by $E = cp$. In relativity theory, such a relationship would be characteristic of a particle with zero rest mass, and such a particle would always automatically travel at the speed of light, c. That the photon does indeed behave, in many contexts, just like a particle with these characteristics is nicely illustrated in the so-called Compton effect (the scattering of light by free electrons); the principal features of this effect can be obtained by simply treating the photon like any other particle and applying the laws of conservation of energy and momentum to its 'collision' with the electron.

Apart from having zero rest mass (and not being deflected by electric or magnetic fields), the photon differs from the other particles we have met so far in two important respects. First, although stable in free space, photons can be created and destroyed very easily in matter; for example, a photon which encounters an atom can raise an electron from a lower-energy to a higher-energy state, transferring to the electron its energy, $E=h\nu$, and disappearing in the process. (By contrast, a single electron cannot just disappear, even in matter, without leaving another particle behind—see below.) Second, it is possible to have a very large number of photons in the same state, and indeed one needs this if one is ever to get back to the 'classical' light wave which gave rise to the idea of photons in the first place. These differences are actually just symptoms of a much more fundamental distinction between photons and these other particles, to which I shall return shortly.

From what I have said so far, one might think that the electrostatic (and magnetic) interactions of charged particles with one another, on the one hand, and their interactions with photons on the other, are two quite different phenomena. In fact, the former can be viewed as a consequence of the latter, and since this feature is a special case of a much more general idea which is central to modern particle physics, we will take a moment to examine it. Consider an electron in free space, and choose, as special relativity permits, a frame of reference in which it is at rest. Then the principle of conservation of energy forbids it to emit a 'real' photon—that is, one which can propagate away to infinity. Suppose, however, that there is another electron a distance r away. Then we can consider the possibility of a process in which a photon is emitted by the first electron and absorbed by the second (Figure 2.1); such a photon is called 'virtual', since it does not appear in the final state of the system. In such a process the photon would carry momentum and energy away from the first electron and transfer it to the second; thus the complete process amounts to a collision, or interaction, between the two electrons. The length of time, Δt, for which the 'virtual' photon must exist is the time necessary to propagate a distance r—that is, r/c. Now, by the 'energy–time indeterminacy relation' mentioned in Chapter 1, if a quantum process lasts for a time Δt, we are allowed to 'borrow' for it an energy ΔE, provided that ΔE is no greater than about $h/4\pi\Delta t$. If the rest mass of the photon were finite, say m, then the minimum value of ΔE would be its rest

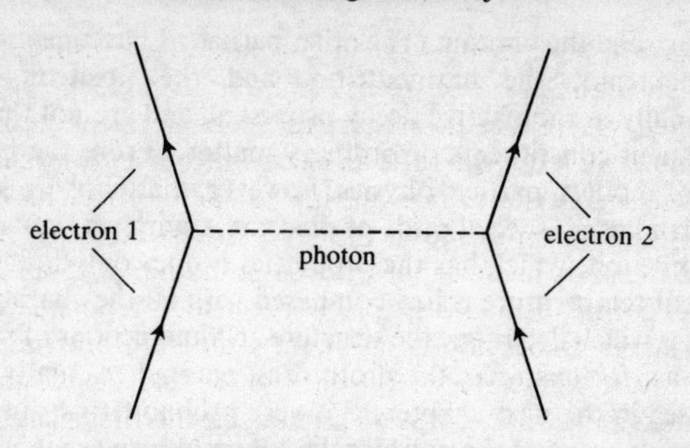

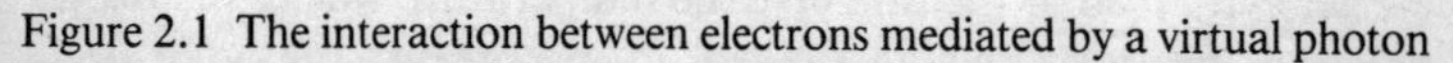
Figure 2.1 The interaction between electrons mediated by a virtual photon

energy, mc^2, and therefore the maximum value of Δt we could tolerate would be of the order h/mc^2, and the maximum value of r correspondingly h/mc. Thus the net result of the whole process would be an effective interaction, or force, between the two electrons which operates only when they are within a distance of less than about h/mc of one another (the so-called 'range' of the interaction). In reality, of course, the rest mass of the photon is zero, so the force thus generated is of infinite range; but it turns out that it falls off as the inverse square of the distance. In fact, a detailed quantum-mechanical calculation shows that such 'virtual-photon exchange' processes indeed generate both the well-known Ampère forces between electric currents and (with some technical provisos which I shall not go into here) also the electrostatic (Coulomb) interactions between static charges. It is not obvious from the above qualitative considerations, but nevertheless turns out to be true, that the Coulomb interaction so generated is repulsive between like charges and attractive between unlike ones, as observed, while the Ampère interaction is attractive between parallel currents and repulsive between antiparallel ones. Thus, one can regard the 'elementary' interaction as being between the charged particle and the photon, and the interaction between the charged particles themselves as being derived from it. The theory constructed in this way is known as 'quantum electrodynamics'.

Thus, most of the properties and behaviour of ordinary matter as we know it here on earth can be explained in terms of four apparently 'elementary' particles—the electron, the proton, the neutron, and the photon. (The other particles I have mentioned—the neutrino, the antineutrino, and the positron—occur principally in radioactive decay processes, and are not therefore permanent constituents of ordinary matter.) From the point of view of modern particle physics, however, matter as we know it on earth has no special pride of place; it is simply a 'low-energy' phenomenon, which has the properties it does only because the ambient temperature is low compared with all the characteristic energies which determine the structure and interactions of particles (such as, for example, the proton rest energy). Actually, as we shall see in the next chapter, it is very probable that during the earlier evolution of the universe the temperature was much higher and the 'particles' which played an important role were quite different from the ones with which we are familiar. In fact, almost all progress in our understanding of the structure of matter at the subnuclear level in the last three decades or so has come from experiments in which the collisions and interactions of particles are studied under conditions in which their characteristic energies are very high compared with the typical energies important in terrestrial phenomena (hence the modern term 'high-energy physics' rather than 'particle physics').

Studying particles and their interactions at such high energies has two main advantages. First, it turns out that there are many particles with rest masses of the same order as, or greater than, that of the proton; being unstable, they are not normally found in ordinary matter, but they can be produced quite readily in high-energy collisions. The study of such particles reveals many clues to the structure of matter which could not have been found by inspecting only the stable particles. Moreover, in some cases a theory of the behaviour of particles currently observed has led ineluctably to the prediction of new particles, often in an energy range somewhat above that currently attainable. Such predictions are of course a major stimulus to the experimental programme, and in a number of cases have been spectacularly verified. Second, crudely speaking, the higher the energy involved in a collision experiment, the shorter the distance scale over which we can investigate the structure and interactions of particles.

To see why this should be so, we first recall that in classical optics the resolution which can be obtained—roughly speaking, the smallest separations which can be distinguished—using light of wavelength λ is of the order of λ itself. In quantum mechanics, as we saw in Chapter 1, particles have a 'wave' aspect, and again it turns out that the smallest scale of structure which can be probed is of the order of the wavelength, λ, which, we recall, is, by de Broglie's relation, inversely proportional to the momentum, p, of the particle: $\lambda = h/p$. Now in the relativistic limit (velocity approaching the speed of light), when the total energy, E, of the particle is large compared with its rest energy, mc^2, E is approximately equal to cp; hence λ is approximately equal to hc/E, so that the higher the energy, the shorter the scale of structure which can be investigated. Thus if, for example, we suspect that the proton, which is known to have a radius of the order of 10^{-15} metres, has some internal structure, and we wish to investigate this by scattering some particle such as an electron from the proton, we need to impart to the electron an energy of at least $hc/(10^{-15})$—that is, about 2×10^{-10} joules, or 3 GeV (1 GeV—giga-electron-volt—is 10^9 electron-volts or 1000 MeV). Similarly, if we wish to investigate the detailed structure of the 'weak' interaction responsible for the decay of the neutron (see below), which is believed to have a range some twenty times smaller than the proton radius, we need an energy of at least around 60 GeV; and so on. Thus, high-energy physics is automatically also the physics of the very small.

Thus, the whole history of experimental elementary particle physics over the last three decades has centred on the attainment of ever higher energies for the colliding particles. Until relatively recently, most experiments used a stationary target—for example, the protons in a bubble chamber—and accelerated a beam of (say) protons or electrons so as to collide with the target protons. Unfortunately it turns out that as soon as we get to energies comparable to the rest energies involved, most of the energy imparted to the beam goes to waste; it simply carries the whole complex (beam particle plus target proton) forwards, whereas what actually matters is the energy of *relative* motion. (If a 20-ton truck moving at 30 m.p.h. hits a stationary VW, the damage done is the same—at least in an ideal world where brake friction and so on can be neglected—as if the truck had been stationary and the VW moving at 30 m.p.h., even though the kinetic energy in the first case is much greater than in the second: in the first case, most of the energy of the truck simply goes into carrying both vehicles forwards.) Were Newtonian mechanics universally valid, this problem would not be too severe, since a constant fraction of the energy originally imparted, at least, would be available for the relative motion (as in the analogy just given). But unfortunately, special relativity makes matters much worse: the available energy increases only as the square root of that originally provided. For this reason, most recent ultra-high-energy accelerators have been of the type known as 'colliders'; rather than accelerating a single beam of particles to hit a stationary target, one accelerates two beams to equal energies but in opposite directions and allows them to collide. In this way all the energy initially provided is genuinely available for the collision process. The disadvantage, of course, is that since the number of particles one can accelerate in each beam is very limited, the number of collisions per second is much smaller than in a fixed-target arrangement; but this price is generally felt to be worth paying. At the time of writing the maximum energy available in existing colliders is about 2 TeV (2×10^{12} electron volts), or about two thousand times the rest energy of the proton, in a ring which is a few kilometres in diameter, at the Fermi National Accelerator Laboratory near Chicago. The construction cost of such an accelerator is typically a few hundred million pounds, and its power consumption in operation about 50 megawatts (about 0.1 per cent of the total electrical power consumption of the United Kingdom). A projected accelerator currently under active discussion in the United States, the so-called Superconducting Supercollider, will if built scale up each of these numbers by a factor of ten to twenty. Just why many people feel it is worth spending as much as four billion dollars on such a machine is something I will come back to later.

Although the technology involved in a typical experiment in contemporary high-energy physics is very sophisticated, the conceptual structure of such an experiment is simple in the extreme. In effect, one simply collides a beam of particles of type A with one of type B and watches what comes out. To be a bit more specific, according to current ideas a particle of a given type, say an electron or a proton, is characterized by at most three variables, which we can take as its direction of propagation, its energy, and an appropriate component of its intrinsic angular momentum, or spin. Thus, if we have prepared the beams of particles A and B to have well-defined values of these three variables, the initial state of the colliding particles is completely defined. (In practice, one often uses 'unpolarized' beams, for which only the direction and energy are controlled; in the theoretical analysis one must then average over the different spin states.) Similarly, if a particle of a given type is produced in the collision, there are again only three pieces of information which we need to describe it: its energy, its spin component, and the direction in which it emerges. Thus, the maximum information we can extract from a given experiment is expressed in the form of a list of so-called *differential cross-sections*, which are the answer to the questions: If I smash a beam of particles of type A into a beam of type B with a given energy (and possibly a given spin) and set my detectors to catch what comes out at a given angle relative to the beam direction, how many particles of type C, D, E . . . with a given energy and spin do I get (including of course A and B themselves)? And what are the correlations between them (for example, what is the probability that a particle of type C emerges at an angle of 20° with an energy of 2 GeV, given that one of type D emerged at −10° with an energy of 800 MeV)? This list of numbers—the magnitudes of the cross-sections—supplemented by a very few other pieces of data from atomic physics and elsewhere, forms the whole of the experimental data base for current theories of the basic structure of matter. (Needless to say, in many experiments one does not extract the maximum possible information; for example, for practical reasons one may be unable to measure the spin of the emerging particles.)

Every now and again it is reported in the popular press that a 'new particle' has been discovered. What does this really mean? How do experimental high-energy physicists know that they have seen a new type of particle? In the simplest case, when the particle is both charged and long-lived, it will ionize the atoms in the detector and leave a track which is visible in a photograph of the event (or, nowadays, is recorded automatically by electronic equipment). Detailed measurements of the properties of the track, such as the ionization density, the curvature in a magnetic field, and so on, may enable the experimenters to deduce the mass and charge of the particle in question, and the distribution of the 'decay products' when it disintegrates often allows them to infer also its spin. (Such an inference requires the study of a number of different decay events.) If it is uncharged, we must proceed more indirectly. In many cases, if the particle decays into charged particles with a 'reasonable' lifetime—say, 10^{-10} seconds, which is typical of decays brought about by the weak interaction (see below)—then, although we cannot 'see' the particle itself, we can see the tracks produced by its decay products and from them infer its properties. (Alternatively, if it makes reasonably frequent collisions with nuclei, as does the neutron, we can track it by the effects of these.) Another way of inferring that a new type of particle has been produced in a collision is to measure the total energy, E, and momentum, p, of the observed 'collision products' (outgoing particles) and check whether they add up to the energy and momentum of the original colliding beams. If not, and if the 'missing' energy and momentum, E_{m} and p_{m}, always satisfy the relation $E_{\mathrm{m}}^2 - c^2 p_{\mathrm{m}}^2 =$ a constant, then we may reasonably infer that there is also produced in the collision a single unobserved particle whose mass is c^{-4} times the constant; the energy and momentum carried off by this particle then make up the deficit. This method relies heavily, of course, on the principles of conservation of energy and momentum, which are assumed to apply universally and in detail to all collisions. It was in this way that the neutrino was 'discovered' in the thirties, and most physicists were quite happy to accept its reality, even though it was another twenty years before, with the help of the intense neutrino flux produced in a nuclear reactor, its effects were directly detected. Some of the more interesting particles discovered in recent years still have the somewhat equivocal status possessed by the neutrino in the thirties and forties.

Many of the so-called particles whose discovery has made recent headlines do not have even this status, however, but were actually identified only in the guise of 'resonances'. To explain the idea of a resonance, let us digress for a moment to atomic spectroscopy, and consider what happens when we shine a beam of light (photons) on a gas of atoms and measure the differential scattering cross-section—that is, the number of photons scattered through a given angle, as a function of the light frequency, ν, or, equivalently, the photon energy, $E = h\nu$. As we know, the electrons in an atom can occupy only certain fixed, discrete energy levels. Suppose that at the beginning of the experiment they are all in the 'ground state' (lowest-energy state), and that the next lowest

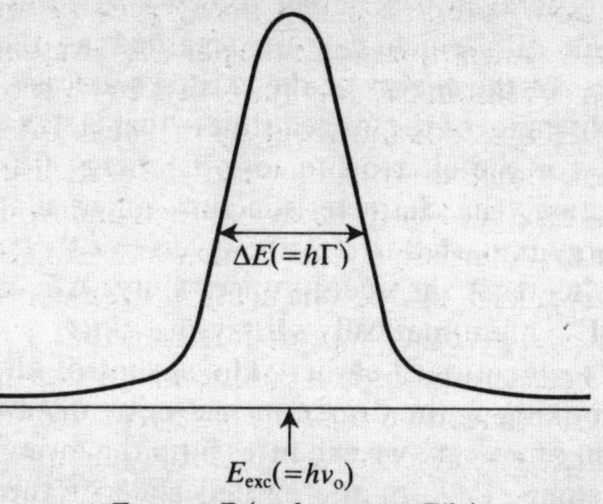

Figure 2.2 The behaviour of the cross-section for scattering of a photon by an atom as a function of photon energy (or frequency)

(the first 'excited') state has an energy which is greater by an amount E_{exc}. If the photon energy, $h\nu$, is not particularly close to E_{exc}, the scattering of the light is weak and shows no very spectacular frequency dependence. But as $h\nu$ approaches E_{exc}, the scattering increases enormously, reaching a maximum when $h\nu$ is exactly equal to E_{exc}, and then drops off again as $h\nu$ moves away from this value. If we plot the scattering cross-section as a function of photon energy, we find a characteristic peak centred at $E_{exc} = h\nu_0$, with some width which we label $h\Gamma$ (see Figure 2.2); this is called a 'resonance'. If we wished for a classical picture of what is going on (as people did in the late nineteenth century, before quantum mechanics), we would say that the electrons behaved like classical oscillators with a resonance frequency $\nu_0 = E_{exc}/h$, and that when the light frequency, ν, is close to ν_0, the oscillators are 'resonantly' excited and can absorb, and then re-radiate, a great deal of power. The width, Γ, of the resonance peak would be a measure of the damping of the oscillators (due to the re-radiation process itself). In the quantum picture, when the photon energy, $h\nu$, is close to E_{exc}, the photon can be absorbed, thereby promoting the electrons from the ground state to the excited state; subsequently the excited state decays—that is, the electrons drop back to the ground state, re-emitting the photon, in general in a different direction. In this picture the quantity $\Delta E = h\Gamma$, which can be regarded as the degree of indefiniteness of the energy of the excited state, is a measure of the inverse 'lifetime' of the excited state—that is, the time it takes, on average, for the electron to lose its energy (just as, in the classical picture, the characteristic time an oscillator takes to lose the energy imparted to it is the inverse of the damping, Γ). In fact, we see that the energy uncertainty, ΔE, and the lifetime, $\Delta t(\sim\Gamma^{-1})$ automatically satisfy the 'uncertainty relation' $\Delta E \cdot \Delta t \geqslant h/4\pi$ mentioned earlier. The upshot of all this is that, merely by looking at the degree of scattering of photons (light) as a function of energy, we can infer both the mean energy, E_{exc} and the lifetime, Γ^{-1}, of the excited state of the atom. It is convenient (if perhaps a little unnatural in an atomic context) to regard the excited state as an unstable 'particle' in its own right; this point of view makes some sense provided the energy is reasonably well defined ($\Delta E \ll E_{exc}$), which for atoms is always the case.

Returning now to particle physics, we proceed in an exactly analogous way. Suppose we observe the scattering of particle A by particle B as a function of total energy, E (including the rest energy of A and B), and that we notice that near some energy E_0 the scattering increases sharply, corresponding in Figure 2.2 to a peak with width ΔE. Then we infer the existence of a 'particle' with an energy E_0, hence a mass E_0/c^2 and a lifetime $h/\Delta E$. Unlike the case of light-scattering by an atom, the particles A and B enter here on the same footing, so it is not particularly natural to regard the new particle as an excited state of either A or B. In many cases it is not even natural to regard it as an excited state of the complex A + B, because it often turns out that precisely at the energy of the resonance there is a large probability also of the process A + B → C + D, where C and D are different from A and B. (For example, in the scattering of negative pions by protons there is a sharp resonance at a centre-of-mass energy of about 1230 MeV, and just at that point one also finds a large probability of the system converting to a neutral pion and a neutron.) So it seems more natural to think of the 'resonance' as a 'particle' in its own right, one that can decay by various modes (into A + B or C + D, or perhaps something else). The resonances commonly observed in accelerator experiments typically have widths ΔE of

~100 MeV, corresponding to lifetimes of the order of 10^{-23} seconds; since in such a time they could travel, even at the velocity of light, only a small fraction of an atomic radius, it is clear that they cannot be observable by their tracks in a bubble chamber or elsewhere; and indeed there comes a point (when ΔE becomes comparable to E_0) where one really has to question the meaning of thinking of them as particles at all. Fortunately, some of the resonances most crucial to the current theoretical framework in fact have lifetimes considerably longer than 10^{-23} seconds (though still not long enough to leave an observable track).

The count of particles discovered in recent years now runs into several hundreds, and there would be no point in trying to list or describe them all here, particularly as most of them are no longer regarded as truly 'elementary'. It is more useful to review the main principles of classification of particles; these rely on a combination of experimental data and theoretical considerations drawn from quantum field theory, the conceptual framework which is used to describe matter at this level (see below).

A given 'particle' is characterized by a number of invariants—that is, quantities which are independent of the details of its motion. One such invariant is, obviously, its mass; at our present level of understanding, it appears that this can be essentially anything. Most of the other invariant quantities, however, seem to take only discrete values—usually integers or simple fractions—and are referred to as 'quantum numbers'. Some of these quantum numbers, such as the spin (the magnitude of the intrinsic angular momentum which the particle possesses when at rest, measured in the natural units of $h/2\pi$),[6] are by their nature positive, while others, such as the electric charge, can be positive, negative, or zero. A very general prediction of quantum field theory, which has been widely confirmed by experiment, is that to every particle there corresponds an 'antiparticle'—that is, a particle with the same value of, for example, mass and spin, but opposite values of quantum numbers such as electric charge. Thus, for example, as mentioned above, in some radioactive decays the decay products include a particle, called for historical reasons the 'positron' rather than the anti-electron, which has the same mass and the same spin ($\frac{1}{2}$) as the electron, but a charge of $+e$ rather than $-e$. Similarly, to the proton there corresponds the antiproton, and to the neutron the antineutron. Even the massless and barely observable neutrino has its antiparticle, the antineutrino. A very few particles, such as the photon, which have no 'reversible' quantum numbers, are in effect their own antiparticles. Crudely speaking, the production of an antiparticle is equivalent to the absorption of the corresponding particle, and vice versa, except, of course, that in the equation of conservation of energy the rest mass appears on the other side. Thus, for example, if in a given radioactive nucleus a proton, p, can decay into a neutron, n, a positron, e^+, and a neutrino, ν—($p \rightarrow n\, e^+\, \nu$)—it can also decay by *absorbing* an atomic electron, e^-, and producing only a neutron and a neutrino—($pe^- \rightarrow n\nu$)—the so-called K-capture process. Similarly, since in appropriate circumstances an electron, e^-, can absorb a photon, γ—($e^-\gamma \rightarrow e^-$)—it follows that, given an energy greater than twice the electron rest energy, the photon can create an electron–positron pair, disappearing in the process—$\gamma \rightarrow e^+e^-$, the so-called pair creation phenomenon. (Because of the need to conserve momentum as well as energy, this process can occur only in matter, not in free space.)

Probably the most important distinction to be made among particles is that between 'bosons' and 'fermions'. According to quantum field theory, the spin of a particle (see above) can take only integral or half-odd-integral values ($0, \frac{1}{2}, 1, \frac{3}{2} \ldots$). Moreover, an assembly of identical particles of a given type is predicted to behave in a totally different way depending on whether the spin is half-odd-integral ($\frac{1}{2}, \frac{3}{2}, \ldots$) or integral ($0, 1, 2 \ldots$). In the former case, it turns out that no more than one particle can occupy a given state; such particles are said to obey 'Fermi–Dirac statistics' and are called 'fermions'. The proton, neutron, electron, and neutrino (which all have spin $\frac{1}{2}$) are all fermions. In the case of the electron, the veto on double occupancy of a state (the so-called Pauli exclusion principle) is crucial to the way in which the electronic structure of the atom is built up, and hence to all of chemistry; similar considerations for the proton and neutron determine the structure of complex nuclei. By contrast, particles with integral spin can have any number occupying the same state, and indeed it turns out that in some sense they prefer multiple occupancy; such particles are said to obey 'Bose–Einstein statistics' and are called 'bosons'. The only 'elementary' boson met with in everyday life is the photon, which has spin 1; it is no accident that this is the only one of the observed stable elementary particles which is associated with a classical wave (light); indeed, it is precisely because the Bose–Einstein statistics allow us to have very large numbers of photons in a single state that we can produce, under appropriate conditions, a classical light wave. Other bosons, observed and/or predicted, include a large number of unstable particles and resonances, the 'graviton' (the predicted spin-2 quantum of the so far unobserved classical gravitational wave—see Chapter 3), the 'gauge bosons' (Ws, Zs, and gluons) to be introduced below, and complexes of even numbers of fermions such as the ^{4}He atom (see Chapter 4). The theorem that particles of half-integral spin obey Fermi–Dirac statistics and those of integral spin Bose–Einstein statistics—the so-called spin-statistics theorem—follows from quite abstract-looking considerations of quantum field theory, and it is very tempting to try to put some flesh on it by visualizing fermions as (say) some kind of dislocation of space–time such that two successively applied dislocations would cancel out; but, so far at least, no totally coherent picture has emerged in this way, and most physicists prefer to stick strictly to the formalism of quantum field theory without trying to interpret it.

A second important distinction is between 'leptons' and 'hadrons'. The vast majority of particles discovered so far are susceptible to the 'strong' interaction which binds the nucleus together (see above), and are called 'hadrons'; if fermions (for example, the proton or neutron), they are called 'baryons'; if bosons, 'mesons'. Apart from the photon and some recently discovered cousins to be introduced below, which are usually put in a class by themselves, there are only six known particles (plus their antiparticles) which do *not* experience the strong interaction. The only two which occur in 'everyday life' are the electron and its neutrino (and their antiparticles). However, cosmic-ray and accelerator experiments have revealed the existence of two other pairs. First, there is the mu-meson, or muon, a particle which appears to be virtually identical to the electron in everything except its mass (which is about 106 MeV—that is, about 200 times that of the electron); the muon turns out to have its own neutrino, the so-called muon, or μ, neutrino, which is not identical with the ordinary (electron) neutrino, as is shown by the fact that it cannot induce the same reactions. Each of the couples (e^-, ν_e)

and (μ^-, ν_μ) is called a 'generation'. Recently a third generation was discovered in accelerator experiments—the so-called tau lepton, τ (mass 1784 MeV), and its neutrino, ν_τ. Whether or not there are yet more lepton generations—perhaps even an infinite number—waiting to be discovered is at present unclear.

Let us now turn to the interactions between elementary particles. At present, it is generally believed that from a phenomenological point of view there are four, or perhaps five, types of 'fundamental' interaction. The one most important at the level of atoms and molecules is of course the *electromagnetic* interaction. This operates between charged particles only (being repulsive for like charges and attractive for unlike ones), is long-range (both electric and magnetic forces fall off only as the inverse square of the distance), and is conceived of as due to the exchange of virtual photons in the way described above. However, as mentioned earlier, electromagnetic forces could not hold the nucleus together—in fact, they would tend to blow it apart, since the protons repel one another—and for this we need the so-called strong interaction. As we mentioned, this is much stronger than the electromagnetic interaction, but is of short range, its strength falling off exponentially with distance, rather than as the inverse square. At the level of the particles we have discussed so far, it is always attractive. The action of the strong interaction is not confined to the stable nucleons (the proton and the neutron), but is shared by the large class of particles called 'hadrons'. A third type of interaction shows up only in radioactive decays and some other processes which would otherwise be forbidden for reasons of symmetry: it is the so-called weak interaction. At the phenomenological level it appears to be of even shorter range than the strong interaction. There is a class of processes which violate even more symmetry constraints than the usual weak-interaction processes (see below); this is usually regarded as a subclass of the effects of the weak interaction, but sometimes as evidence for yet another 'fundamental' interaction, the 'superweak' interaction. Finally, whatever other forces particles are or are not subject to, as far as we know, they all experience the effects of the gravitational interaction. This resembles the electromagnetic interaction in being long-range (like the electromagnetic force, the gravitational force falls off as the inverse square of the distance), but differs from it in that it is universally attractive (with a strength which is proportional to the product of the masses involved). If we express the strengths of the various interactions in what turn out to be the 'natural' units, then the order of magnitude is very roughly as follows: strong interaction, 1; electromagnetic, 10^{-2}; weak, 10^{-13}; superweak, 10^{-16}; gravitational, 10^{-39}. Thus, ironically, the force which most obviously dominates our everyday lives, gravitation, is at the level of elementary particles by far the weakest of all known interactions!

At this point it is amusing to make a small historical digression. Most physicists are confident that the four interactions (excluding the superweak) listed above are the *only* 'fundamental' interactions, and that when coupled with the data of cosmology they can 'in principle' explain all known natural phenomena. On what is this confidence based? In particular, why are we so sure that the *only* long-range force between electrically neutral bodies is gravitation? Could there not, for example, be a long-range force which acts differently on protons and neutrons, and hence is material-dependent? From an experimental point of view, this question is equivalent to the question of whether the ratio of (apparent) gravitational mass to inertial mass (see pp. 97–8) is identical for all bodies, whatever their nature. This question was addressed in a famous series of experiments by the Hungarian physicist Eötvös and his collaborators, the results of which were published after his death in 1922. The stated conclusion of their paper is that within the accuracy of the experiment (about 1 part in 10^9) the ratio of gravitational to inertial mass is indeed independent of the material used. This conclusion was exactly what was expected in 1922 in the light of Einstein's general theory of relativity (and indeed from more general considerations). For sixty years these results stood unchallenged; although related experiments have been done, the precise experiment done by Eötvös has apparently never been repeated in quite its original form. In early 1986, however, a paper appeared in which the authors, *simply by re-analysing Eötvös's original data*, found that there was in fact a systematic deviation of the ratio from material to material which lay well outside his quoted error, and which is consistent with the hypothesis of an intermediate-range force which acts differently on protons and neutrons. Whether the effect is a real one, or whether it is due, for example, to some systematic and unaccounted-for error in the original experiment, it is certainly much too early to say; however that may be, the moral is that it is all too easy to ignore or brush aside small anomalies which at the time seem to have no theoretical significance (compare the case of the precession of the perihelion of Mercury, discussed in the next chapter). If the postulated 'fifth force' should turn out to be genuine, no doubt a whole host of phenomena will rapidly be discovered which cannot be explained by the standard four fundamental interactions, and which, with hindsight, we must have been blind not to have noticed!

In trying to make sense of the bewildering variety of observed particles, and the ways in which they interact and decay into one another, it is almost impossible to overemphasize the role played by 'symmetry' (or 'invariance') principles and the related conservation laws. Most of us are familiar, in the context of classical physics, with the idea that the laws of physics do not depend on where we are in space, or on the time at which we do a particular experiment. Technically, we say that the laws are 'invariant under translation in space and time'. Already in classical mechanics, these invariance principles lead to conservation laws: invariance under time translation implies that the total energy of an isolated system of particles is conserved (does not change in time), and invariance under space translation that its total momentum is similarly conserved. (Thus, for example, when a rocket expels gas, the total momentum of the whole system, gas plus rocket, remains constant, so the rocket is accelerated forwards.) From the point of view of special relativity, in which there is a close connection between space and time translations, conservation of energy and momentum are just two aspects of the same phenomenon. These conservation laws are among the most fundamental bedrocks of modern physics, and any experiment which appeared to violate them would certainly be subjected to very sceptical scrutiny indeed.

Another very important invariance principle which already appears in classical mechanics is that of 'isotropy', invariance under rotation—that is, for an isolated system all directions in space appear to be equivalent. As far as we know at present, this principle applies not merely in the laboratory, but on the scale of the universe as a whole (see next chapter). This too leads to a conservation law which is believed to be quite universal, that of the total angular momentum of an isolated system of particles. When we come to quantum mechanics, there is an important difference between the effects of invariance under space–time translation (in infinite space) and under rotation, which essentially arises because if we continue to rotate a body in the same direction, we eventually come back to where we started: namely, the conserved quantities in the first case (energy and momentum) can even in quantum mechanics take any values, whereas the invariant corresponding to rotation—namely, angular momentum—is quantized:[7] in units of $h/2\pi$ it can take only integral or half-integral values. More precisely, since angular momentum is a vector quantity, the correct statement is that its magnitude can take, in these units, only the value $\sqrt{J(J+1)} \times h/2\pi$, where J is integral or half-integral, and its projection on any given axis can then take any of the values J, $J-1$ . . . $-J+1$, $-J$. For an isolated particle at rest, the value of J is constant and characteristic of the particle: as mentioned above, it is usually called the 'spin'. In some sense it can indeed be thought of as due to an intrinsic spinning motion of the particle; but this picture, if taken too literally, soon leads to serious difficulties, and it is probably better not to put too much weight on it. A point which will be very important when we consider more abstract symmetries is that, from a purely formal point of view, it is possible to construct a particle of arbitrary integral or half-integral spin by combining an appropriate number of particles of spin $\frac{1}{2}$. For example, if we take two particles of spin $\frac{1}{2}$ and put them together in such a way that the projection of each spin on a given axis (say the z axis) is $+\frac{1}{2}$, then not only is the resultant spin projection on that axis $+1$, as it obviously must be, but, less obviously, the magnitude of total angular momentum so generated is $\sqrt{2}$ as required for a $J=1$ particle. If, on the other hand, we combine the same two particles of spin $\frac{1}{2}$ so that their projections on the z axis are opposite, we can get a 'particle' with both projection and total magnitude of spin equal to 0—that is, a $J=0$ particle. 'Particles' with spin $\frac{3}{2}$, 2, . . . can be built up similarly. The two possible states with total spin $J=\frac{1}{2}$ and projections $+\frac{1}{2}$ and $-\frac{1}{2}$ respectively can therefore be regarded as the basic building blocks out of which higher-spin (and spin-zero) states can be constructed. Note, for future reference, that a rotation of the coordinate system around the z axis leaves our two basic building blocks unchanged, whereas a rotation of 180° around any axis in the *xy* plane will interchange them; other rotations turn out to have a rather more complicated effect, producing from either one of the building blocks what in quantum mechanics is called a 'linear combination' of the two.

Another concept which it is useful to introduce at this stage is that of 'broken symmetry'. As an example, we note that, although I stated above that for a truly isolated system all directions in space are equivalent, when we are doing physics in a terrestrial laboratory, it is much more natural to distinguish the vertical direction from the others; so long as we do not explicitly include the earth in our calculations, the laws of motion of, say, a cricket-ball are certainly *not* invariant under rotation! We say that the original symmetry (invariance) under three-dimensional rotation has been 'broken' by an external factor—namely, the gravitational field of the earth. In this case the breaking is only partial, since even for a cricket-ball the laws of motion are still invariant under rotation around a vertical axis. As a result, regarding the ball as an isolated system, we can say that the vertical component of its angular momentum, but not the horizontal components, is still conserved.

In the above example the original isotropy of the laws of motion was destroyed by an external agency; this is called 'externally broken symmetry'. There is, however, another kind of broken symmetry, which has played a very important role in particle physics in recent years—namely, what is called 'spontaneously broken symmetry'. This concept arises most naturally in the context of the large assemblies of particles dealt with in condensed-matter physics, and at this point I shall have to anticipate some considerations which will be explained in more detail in Chapter 4. Let us consider a very idealized model of a magnetic material—a model in which the electrons are visualized as tiny magnets sited at points in a lattice, and interacting with one another by forces which are completely isotropic. Thus, in considering the interaction energy (mutual potential energy) of a particular pair of electronic 'magnets', we suppose that it is lowest when they are oriented parallel to one another, but that it does not depend on the direction in space in which they both point. It is clear that in this model all directions in space are completely equivalent, and the laws of motion of the magnets are invariant under rotation. If the system were to possess a net magnetization, it would have to be oriented in a particular direction, and at first sight this would seem to violate the principle of isotropy. Nevertheless, at sufficiently low temperatures in this model, that is exactly what the system does; the electronic magnets all begin to point in the same direction, thereby generating a macroscopic magnetization in that direction. Thus, the laws of motion remain invariant under rotation, but the equilibrium state of the system violates the invariance. To be sure, the real situation is a bit more complicated; in any remotely realistic case the direction of magnetization chosen will in fact be determined by small random magnetic fields which will destroy the rigorous rotational symmetry, thereby giving rise to external symmetry breaking; however, the important point in the present context is that, by making the size of the system larger and larger, we arrive at a situation in which it needs a smaller and smaller field to pick out the direction. Thus, it is tempting to argue (though there are some delicate conceptual problems here!) that in the limit of an infinite system the spontaneous symmetry breaking will occur even in the complete absence of an external field, the direction chosen then being quite arbitrary.

The symmetry properties we have considered so far refer to continuous changes—translations and rotations—of the space-time reference frame, and as far as we know are universally valid. However, there are two discontinuous changes we can consider: namely, the operation of going from a right-handed to a left-handed set of coordinate axes (known technically as the operation P, for 'parity'), and that of reversing the direction of the time axis, T. Are the laws of physics invariant under P and T? That is, can Nature tell her right hand from her left, or the 'forwards' direction of time from the 'backwards' one? At first sight it is extremely tempting to argue that these distinctions, particularly the right-left one, have a purely anthropomorphic significance, and that 'obviously' Nature cannot tell the difference. That was indeed the view which most physicists regarded as 'common sense' until 1956—in which year it was suggested on theoretical grounds, and subsequently verified experimentally, that while the strong, electromagnetic, and (presumably) gravitational interactions are indeed invariant under both P and T, the weak interaction is not. Indeed, the weak interaction violates P 'maximally'; for example, if a neutrino is emitted in a radioactive decay, it is always polarized 'left-handedly' relative to its direction of motion—that is, as viewed along the direction of motion, it always appears to spin anticlockwise. (As far as is presently known, 'right-handed' neutrinos do not exist in nature, although this is still an open question.) However, if simultaneously with the P operation, we perform the operation of charge conjugation, C—that is, change every particle into its antiparticle—the symmetry is restored; for example, the antineutrino is always 'right-handed'. Thus, the interaction responsible for ordinary radioactive decay of nuclei is invariant under the *combined* operation CP, though not under C or P separately. Unfortunately, the picture is further confused by the fact that there is a small subclass of (apparently) weak-interaction-induced decay processes which are not even invariant under the combined operation CP. These are sometimes attributed to a 'superweak' interaction, as mentioned above. There is a very fundamental theorem of quantum field theory, the so-called CPT theorem, which has so far stood the test of experiment, that *all* interactions must be invariant under the simultaneous application of C, P, and T; thus, the superweak interaction is presumably also not invariant under T alone. This 'CP-violation' phenomenon is at present very poorly understood.

All the symmetry operations we have discussed so far, except for charge conjugation, refer to changes in the space–time reference frame. However, there is another class of symmetry operations—the so-called 'internal' symmetries—which has been extremely important in classifying and grouping the elementary particles, and which, as far as is known, has no relation to space–time structure. Historically, the idea arose from the observation that not only do the neutron and the proton have the same mass to within 1 part in 600, but once one has allowed for the purely electromagnetic effects which are due to the proton charge, and so on, their interactions are also almost identical (for example, the data of nuclear physics show that the interaction of two protons is almost identical with that of two neutrons, once the electromagnetic effects are subtracted). Thus it seemed natural to view the neutron and the proton as two states of a single entity, the nucleon, which are related in the same way as the two states of a spin-$\frac{1}{2}$ particle corresponding to projection $+\frac{1}{2}$ and $-\frac{1}{2}$ on the z axis, and to postulate the existence of a corresponding abstract 'space' ('isotopic spin space') such that the strong interaction, at least, is invariant under arbitrary rotations in this space. As a result, *all* components of the 'isotopic spin' are conserved by the strong interaction. This hypothesis, when applied not just to the proton and the neutron but to the other (now) known hadrons, immediately has very fruitful consequences: not only do many of the particles fall into rather simple patterns ('multiplets', that is, groups of particles of the same (real) spin and nearly the same mass), but we can frequently understand the relationships of the observed scattering cross-sections. Isotopic spin is believed to be conserved by the strong interaction, but not by the electromagnetic or weak interactions; the latter may be said to break the symmetry, rather as the earth's gravitational field breaks the rotational symmetry in ordinary three-dimensional space. This has the consequence that when the total isotopic spin of the decay products of a given particle is different from that of the particle itself, its decay cannot be induced by the strong interaction but must be due to the electromagnetic or weak interaction; since the lifetime is, crudely speaking, inversely proportional to the strength of the interaction responsible for the decay, this in turn means that the particle is considerably longer-lived than might have been expected.

As more and more particles were discovered and their interactions and decays studied, however, it became clear that isotopic spin symmetry was not the whole story. There had to be at least one more quantum number, called for historical reasons 'strangeness', which is conserved by the strong and electromagnetic interactions but not by the weak interaction. It is only thanks to the conservation of strangeness that we see so many unstable particles leaving spectacular tracks in detectors such as bubble chambers; were it not for this principle, such particles would decay by the strong interaction (with a lifetime typically $\sim 10^{-23}$ seconds) or at best by the electromagnetic interaction ($\sim 10^{-19}$ seconds), and hence be observable only as resonances, rather than decaying, as in fact they do, only by the weak interaction with a lifetime $\sim 10^{-10}$ seconds.

In the early sixties it was realized that a much more satisfying picture emerged if one regarded conservation of isotopic spin and of strangeness as only special cases of invariance under a much more general abstract symmetry operation, described technically by the symmetry group known to mathematicians as SU(3). To explain this (in so far as it can be explained without going into technical details) it is convenient to shift our point of view on the question of ordinary rotation symmetry somewhat. Rather than regarding the operation of rotation of the space axes as fundamental and the two building blocks—that is, the states of total spin $\frac{1}{2}$ and spin projection $\pm\frac{1}{2}$—as derived from this, it is more helpful to regard the building blocks themselves as the fundamental quantities. Indeed, with some technical reservations I shall not go into, the group of rotations in three-dimensional space is equivalent to the abstract group of operations known to mathematicians as SU(2), which is essentially the set of operations which transform two basic objects, complex in the mathematical sense, into 'linear combinations' of one another. Similarly, SU(3) is the set of operations which transform three complex basic building blocks into one another. It was realized that, although the observed elementary particles could not correspond to the basic building blocks themselves, most of the observed hadrons could be built up, formally at least, by combining these blocks and the blocks which, if they represented real particles, would correspond to their antiparticles, and many of the observed patterns of the particles themselves and also of their interactions and decays were readily understandable in this way. For example, if the three building blocks are called u, d, and s (for 'up', 'down', and 'strange') and their 'anti-blocks' $\bar{u}$, $\bar{d}$, and $\bar{s}$, then the proton can be constructed by putting together two u's and one d, and the neutron similarly from two d's and one u, while the (unstable but fairly long-lived) particle known as the positive pion or pi-meson (π^+) is equivalent to one u plus one $\bar{d}$. The s block, as its name implies, is needed as a constituent of the observed 'strange' particles. As well as satisfactorily grouping most of the then known hadrons, the SU(3) classification scheme was able to make some remarkable predictions. In particular, it was observed that there should be a pattern (multiplet) of ten particles, of which nine could be identified with already known resonances. By applying the symmetry scheme, it was possible to predict the properties of the tenth member of the multiplet, including its mass, and to show that it could decay only with a change of strangeness—that is, by the weak interaction; it should thus have a lifetime of the order of 10^{-10} seconds and be observable not as a resonance, but as a genuine 'particle' which could leave a track in a bubble chamber. The subsequent experimental observation of a particle with just the predicted properties—the Ω^-—was a major triumph for the theory. As we shall see, in recent years it has been necessary to add more building blocks—commonly nowadays said to correspond to different 'flavours'—to the original three, to account, *inter alia*, for yet more observed particles; but the basic principle remains the same.

Apart from the ('flavour') SU(3) symmetry which we believe (with its more recent extensions) applies to the strong interactions of the hadrons, various other internal symmetries are thought to apply to the known particles, either universally or as regards some interactions, and to generate corresponding conservation laws. For example, it is possible to view the conservation of electric charge as arising from (unbroken) invariance under rotations around a particular axis in (another!) abstract space. Similar considerations apply to 'electron number', which, at least until very recently, was believed also to be universally conserved. (The electron and neutrino are assigned electron number +1, the positron and antineutrino −1: in all decay processes observed to date the total electron number is indeed conserved.) As we will see, a more sophisticated symmetry has recently been invoked to explain various properties of the weak and electromagnetic interactions.

Before leaving the subject of symmetry, it is necessary to say a word about one other type of symmetry property which has played a major role in high-energy physics in the last two decades: namely, gauge invariance. While from a mathematical point of view it is an extremely elegant concept, gauge invariance is a subtle idea and none too easy to visualize intuitively.[8] Perhaps the easiest way to introduce it is to recall a result very familiar from ordinary classical mechanics. If we choose a fixed, time-independent system of coordinates, then we expect Newton's laws to apply; and in particular, we expect a system subject to no external forces to move with constant velocity. However, we may prefer to choose a coordinate system which is time-dependent, and indeed this is the natural choice if we are doing experiments in an earth-bound laboratory, for the axes to which we most naturally refer our observations rotate with the earth. It is a standard result that we are quite free to use a rotating (that is, time-dependent) coordinate system, *provided* that we introduce simultaneously extra (apparent) forces—namely, the well-known Coriolis and centrifugal forces. Now, what if we want our coordinate system to depend not just on time but also on position in space. Can we do that? Yes, we can, again at the cost of introducing extra forces; the procedure for doing this is included in the formalism of general relativity (which is, in fact, historically the first gauge theory).

Now, all this can be applied not just to choice of an ordinary three-dimensional space coordinate system, but also to the choice of 'axes' which define the *internal* symmetries of the system. Moreover, when one goes over to a quantum-mechanical description, the extra forces appear as extra (boson) fields (see below), which in general will have their own dynamics. The oldest and simplest example of such fields acting on an internal symmetry is electromagnetism. As we have seen, the conservation of electric charge can be regarded as a symmetry with respect to rotation around a particular axis (say the z axis) in an abstract space, and the field which is introduced when we decide to make the choice of x and y axes a function of space and/or time is just the electromagnetic field, or, from a quantum-mechanical point of view, the photon. What is crucial is that the photon has its own dynamics. Thus, if we make the internal axes time- or space-dependent in one region of space, the associated electromagnetic (photon) field will rapidly propagate to neighbouring points and interact with any disturbance of the axes there; in this way the standard electromagnetic interaction of charges and currents is generated.

That electromagnetism could be formulated as a gauge theory had been known for many years, but no special significance was attributed to this, since no particular consequences follow which cannot be obtained in other ways. Fundamentally, this is because the relevant symmetry in this case is simply a rotation around a single axis in the abstract space. Such rotations have the property that the *order* in which they are carried out is irrelevant (a rotation of 60° followed by one of 20° takes us to precisely the same state as a 20° rotation followed by a 60° one); the rotation group is technically called 'abelian', and theories involving only such groups are called abelian (gauge) theories. The full richness and potential of gauge theories only began to emerge when people started to consider *non*-abelian theories—that is, theories in which the result of the operations of the relevant symmetry group does depend on the order in which the operations are performed. The group of all possible rotations in a three-dimensional space, either geometrical or abstract, has this property: for example, it is obvious that the result of rotating an asymmetric object such as a book first by 90° around a vertical axis and then by 90° around a horizontal axis is different from that of the two operations performed in the reverse order. Non-abelian gauge theories are very much more difficult to analyze than the abelian variety, and turn out to have some quite unexpected consequences (see below). Fundamentally, this is because, unlike the situation in an abelian theory, the 'gauge bosons'—the fields which have to be introduced—must partake of the properties of the original objects. Thus, while in electromagnetism (an abelian theory) the electron has an electric charge but the photon does not, in a theory in which the relevant symmetry is the full isotopic spin rotation group, the gauge bosons must in general themselves carry isotopic spin. The idea of a gauge theory has proved so fertile that, nowadays, there are very few seriously considered theories of the interaction of elementary particles which are *not* gauge theories.

So far I have said nothing about how a particle theorist would actually go about calculating anything quantitatively. Actually, the standard conceptual framework for such calculations—Lagrangian quantum field theory—is simple in the extreme, so let me try to give at least a very crude sketch of how it works. One first assigns to each different type of particle, together with its antiparticle, a 'field'—that is, a quantity (in general, a complex number in the mathematical sense) which at any particular time is given as a function of position in space. (This field, in the general case, has no direct physical significance.) One then characterizes a particular state of the system under study at a given time—let us say, a state corresponding to firing a beam of particles of type A at one of type B in a given direction with a given energy—not by specifying a unique 'field configuration'—that is, the value of the field at every point in space—but rather by giving a 'probability amplitude' that each such configuration will occur. (This is a generalization of the idea of a probability amplitude used in elementary one-particle quantum mechanics; see Chapter 1.) Roughly speaking, the set of probability amplitudes for the field to have a certain value at a certain point gives us the probability of finding one or more particles of the type in question at that point, and so is related to experimentally measurable quantities like differential cross-sections.

This prescription, when suitably fleshed out, tells us how to *describe* the relevant system of particles. But how do we predict how it will behave? To do this, we need to calculate the change of the probability amplitudes with time, starting from a given initial state such as that described above. The prescription for doing this looks at first sight rather bizarre, and is the product of a long and somewhat tortuous historical development. We simply make a particular choice of a function known as the 'Lagrangian', which depends on the various fields in question and their space and time derivatives. Once this is given, there is a standard and unique procedure for obtaining from it the equations of motion of the system—that is, the way in which the probability amplitudes change with time—and hence for predicting experimental quantities like scattering cross-sections. The Lagrangian cannot be just any arbitrary function, but must satisfy various constraints—for example, those imposed by the requirements of special relativity and the symmetry principles one believes to be obeyed by the interactions of the particles in question (which are, in a sense, built into the Lagrangian); it is this feature which makes the whole procedure something more than just an exercise in data-fitting, and indeed the consequences of imposing these apparently innocent constraints turn out to be amazingly rich. While the actual carrying-out of this programme requires, needless to say, considerable sophistication and technical skill, the basic principle is essentially no more than I have outlined: once one has chosen the nature of the fields and the form of the Lagrangian function, everything is in principle determined, and the rest is calculation. In fact, when a particle theorist says that he has formulated a 'new theory' of such-and-such a phenomenon in high-energy physics, what he means as often as not is simply that he has written down a new Lagrangian, perhaps involving one or more fields corresponding to currently unobserved particles. What he does *not* mean—or means very rarely—is that he has in any way changed the basic conceptual framework of Lagrangian field theory within which he and his colleagues work.

The last two decades of particle physics have been dominated

Table 2.1

	Spin (units of $h/2\pi$)	Electric charge (units of e)	Strangeness
Up quark (u)	$\frac{1}{2}$	$\frac{2}{3}$	0
Down quark (d)	$\frac{1}{2}$	$-\frac{1}{3}$	0
Strange quark (s)	$\frac{1}{2}$	$-\frac{1}{3}$	-1

by two major developments, usually known as 'quantum chromodynamics' and 'electroweak unification'. Quantum chromodynamics is a theory of the strong interaction and the structure of hadrons; electroweak unification, as the name implies, is the derivation of the electromagnetic and weak interactions from a common origin. Both have relied heavily on the idea of gauge theory.

To introduce the ideas of quantum chromodynamics, let us go back to the SU(3) classification of hadrons and ask the obvious question: Do the elements of the fundamental representation correspond to actually existing particles? The mere fact that the observed patterns of hadrons can be *formally* built up by combining these elements does not necessarily mean that they have any real existence—any more than the fact that all allowed angular momentum states can formally be built up by putting together spin-$\frac{1}{2}$ particles necessarily implies that such particles exist in nature. (In fact they do, but as far as is known, there would be no internal contradiction in a world consisting entirely of particles with integer spin.) Nevertheless, it is obviously a very attractive hypothesis that the basic building blocks of the SU(3) group do indeed 'exist' as real particles. Such entities are called 'quarks'; their basic quantum numbers can be readily worked out, and are given in Table 2.1. Most striking is the fact that their charge is an integral multiple (positive or negative) not of the electron charge e, but of $\frac{1}{3}e$. If one takes this picture, then the known baryons are each made up of three quarks, while the mesons are made up of a quark and an antiquark. It can be seen from the table that in each case the charge of the composite particle is an integral multiple of e, as required. For example, the proton is composed of two up quarks and a down one (charge $2(\frac{2}{3}e)+$

$(-\frac{1}{3}e) = e$), while the neutron is two downs and an up $((\frac{2}{3}e) + 2(-\frac{1}{3}e) = 0)$. Although the quarks originally postulated, which are the building blocks of SU(3), are just the three given in Table 2.1, it is now believed that there is certainly at least one more (c, for 'charm'), and almost certainly a further two more (t and b, for 'truth' and 'beauty', or, more prosaically, 'top' and 'bottom'). These new quarks allow the possibility of new baryons and mesons beyond those of the SU(3) classification: in particular, the so-called ψ group of meson resonances first discovered in 1974 are naturally interpreted as combinations of the charmed quark and its antiquark, and the Υ group (1977) similarly as a combination of the bottom quark and its antiquark.

The picture which emerges is a very pleasing one, and is quite in the long tradition of dissecting apparently 'fundamental' entities into yet more fundamental ones. In particular, many of the mass differences within the hadron multiplets can be explained simply by the fact that while the up and the down quarks have very nearly the same mass, the strange quark is appreciably heavier, and the charmed quark very much heavier. At first sight, mesons and baryons are 'composed' of quarks in just the same sense as molecules are composed of atoms and nuclei of nucleons.

There is, alas, one major snag with this picture: no one has ever seen a quark! Whenever in the previous history of physics a composite system has been analyzed into its components, it has always been possible, eventually, to see each of the components in isolation. (For example, the neutron was not only identified as a component of the nucleus, but was 'seen' as an isolated particle, in the sense that one could follow its solo progress from the emitting to the absorbing nucleus. The proton, of course, can be 'seen' much more directly, by the tracks it leaves in a detector.) Now all the quarks have a finite electric charge, and should therefore leave tracks in a detector which are very characteristic (since, *inter alia*, the ionization produced should be characteristic of a charge $\frac{2}{3}e$ or $\frac{1}{3}e$, rather than e). Yet no one has ever reliably seen such a track, and (with one exception which is generally regarded as highly controversial) all other attempts to find free particles with fractional charge have failed. It seems that if quarks really do exist, they exist *only* inside hadrons, never in isolation.

How can this be? Is it just that quarks are very tightly bound together in hadrons, so that the 'ionization energy' necessary to knock them out is very large? Conceivably. But in recent years hadrons have been smashed together at energies of relative motion which are hundreds of times their rest energy, so it would need a very peculiar 'conspiracy' to make this explanation viable. The resolution of the dilemma which has emerged in recent years rests firmly on the hypothesis that the quarks in hadrons interact via a gauge field which is associated with an abstract non-abelian internal symmetry. Although the relevant symmetry group happens to be SU(3), it turns out, somewhat confusingly in the light of the history, that the abstract symmetry in question has nothing to do with the original 'flavour' quantum number (up, down, strange, and so on), but is rather associated with a *new* internal quantum number, which for want of a better name is known as 'colour'. (Hence 'chromodynamics' from the Greek *chroma* meaning 'colour'.) Thus, the original triplet of basic elements (u, d, and s) is replaced for this purpose by a new triplet, say 'red', 'green', and 'blue'. Symmetry transformations in the relevant abstract space change a quark of one colour into a linear combination of three colours; the associated (boson) gauge fields are called 'gluons', and, because the symmetry in question is non-abelian, must themselves carry 'colour'. This feature, as remarked above, fundamentally changes many of the consequences of the theory. In particular, it is generally believed that it leads to the property known as 'confinement'. To explain this, let us consider what happens when, for example, we insert a positively charged impurity into a metal. The relevant interaction in this case is the usual Coulomb interaction, which corresponds to an abelian gauge theory, and what happens is that the free electrons in the metal tend to 'screen' the impurity by building up a cloud of negative charge around it, thereby neutralizing it. Thus, if we try to detect the presence of the impurity by inserting some charged test particle, then at very short distances (well inside the screening cloud) we will indeed see the original impurity charge; however, as we go to large distances, we will see the combined effect of the impurity and the screening cloud—that is, zero net charge. Thus the 'effective charge' of the impurity is large at short distances and small at long distances. As a matter of fact, it is not necessary to have a metallic environment; even in a vacuum we get a similar effect in principle, because of the production of 'virtual' electron–positron pairs (see p. 50) by the photons, themselves virtual, which propagate the interaction, although in real life it is almost negligibly small (that is, one would have to go to almost impossibly short or long distances to see an appreciable variation in the effective charge).

The astonishing thing about non-abelian gauge theories such as quantum chromodynamics is that the effect seems to be totally reversed. That is, although all the details are not yet completely clear, it appears that the gluons, and the associated virtual production of quark–antiquark pairs, actually '*anti*screen' the 'colour charge', in such a way that the effective charge, or the effective quark–quark interaction, is weak at short distances and strong at long ones. In fact, to remove a quark, originally bound in a hadron, 'to infinity' requires not just large, but infinite, energy! Thus, no accelerator, however powerful, will ever be able to produce free quarks, and the only directly observable hadrons which will ever emerge from collisions are three-quark complexes (baryons) or quark–antiquark complexes (mesons). Of course, at *short* distances there is nothing to prevent quarks from changing places; for example, if we shoot a positive pion ($u\bar{d}$) at a neutron (udd), one of the down quarks in the neutron can exchange with the up quark in the pion, giving ($d\bar{d}$) + (uud)—that is, a neutral pion and a proton. Quantitative calculations confirm that this picture is consistent with experiment. Moreover, the idea that the baryons, for example, are 'actually made of' quarks is given strong support by the fact that in high-energy electron scattering experiments the electrons behave just as if they were indeed scattered by three distinct point particles of charge $\frac{2}{3}e$ or $\frac{1}{3}e$. Although not all the consequences of quantum chromodynamics have yet been worked out, the general features it predicts are in such good agreement with experiment that by now very few physicists doubt that it is indeed as good a theory of the strong interactions as quantum electrodynamics is of the electromagnetic ones. The fact that the inter-quark interaction becomes weaker and weaker at short distances, or, what is in effect equivalent, at high energies, is known as 'asymptotic freedom' and is of great significance for cosmology (see Chapter 3).

The second major theme of the last twenty years has been the unification of the various interactions. Recall that there are usually thought to be four fundamental interactions: the gravitational interaction, which acts on everything; the electromagnetic interaction, a long-range interaction between charged particles; the strong interaction, a short-range force acting between hadrons (that is, as we now believe, quarks); and the weak interaction, an ultra-short-range interaction which acts on both quarks and leptons. As always, physicists faced with this kind of situation instinctively ask the question: Are these interactions 'really' different, or could two or more of them be simply different manifestations of the same underlying phenomenon (just as Maxwell showed that the apparently totally different phenomena of electricity and magnetism were in some sense one and the same thing)? Indeed, Einstein spent the last thirty years of his life mainly on an unsuccessful attempt to unify the electromagnetic and gravitational interactions—a highly plausible attempt at first sight, in view of their obvious similarities (both long-range and so on). While Einstein's goal is today almost as far out of sight as ever, it is remarkable that we *have* been able to unify the electromagnetic and weak interactions. At first sight these are just about the most implausible candidates for unification. The electromagnetic interaction is long-range, the weak interaction very short-range; the former conserves parity, while the latter violates it 'maximally'; and so on. Why should anyone even have thought of doing such a thing?

Let us start by recalling an observation which at first sight seems irrelevant. When, in a radioactive decay process, electrons and neutrinos and/or their antiparticles are produced and/or absorbed, the total 'electron number'—that is, the number of electrons plus (electron) neutrinos minus the number of positrons plus antineutrinos—always seems to be conserved. Similarly, in processes involving muons the number of muons plus muon neutrinos minus their antiparticles is conserved. An electron never simply vanishes without leaving a (electron) neutrino (unless it has absorbed an antineutrino), and similarly for the muon. What this suggests is that perhaps the electron and the neutrino can be regarded as two alternative states of some underlying basic entity, related by some symmetry operation (just as in the case of the proton and the neutron). In the light of this idea, let us consider the K-capture process mentioned earlier, in which a proton in a nucleus 'captures' an electron and turns into a neutron, emitting a neutrino in the process. Compare this with the process in which the proton *scatters* an electron. In the latter case, we saw that the

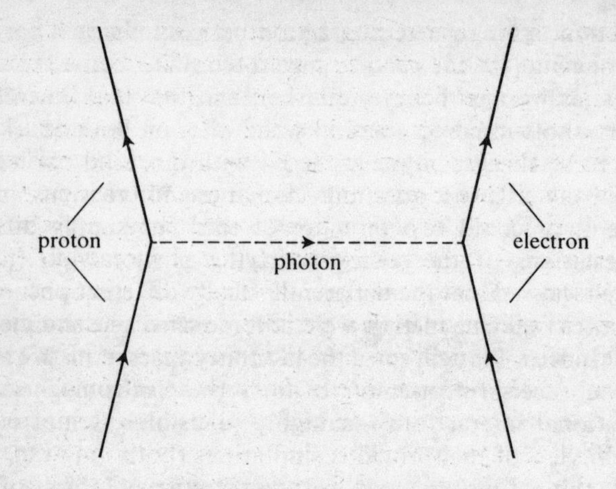

Figure 2.3 The scattering of an electron by a proton, viewed as mediated by exchange of a virtual photon

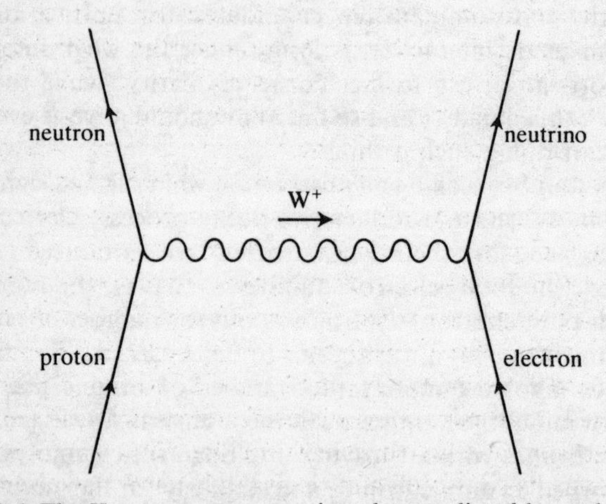

Figure 2.4 The K-capture process, viewed as mediated by exchange of an intermediate vector boson, W^+

process can be represented as the emission of a virtual photon by (say) the proton and its re-absorption by the electron (Figure 2.3). Proceeding by analogy, it is extremely tempting to think of the capture process as also proceeding in two steps (Figure 2.4), involving the emission and re-absorption of a virtual particle of unknown type represented by the wiggly line. What would such a hypothetical particle have to be like? First, to fit into the general gauge-theory picture, it turns out that it would have to be a spin-1 boson, just like the photon. Second, it would have to be charged, since in the K-capture process the left-hand side of the picture loses a unit of charge and the right-hand side gains one. Third, its interactions would have to violate parity conservation—a general characteristic, as we saw, of the weak interaction. To account for all observed radioactive decay processes, one needs both a positively and a negatively charged particle of this type; they are called '(weak) intermediate vector bosons' (IVBs) and are denoted by the symbols W^+ and W^-. But then one is tempted to ask: if there is a positively and a negatively charged particle, should there not also be a neutral one of similar characteristics? The postulation of such a particle (called, for technical reasons, Z^0 rather than W^0) has some remarkable consequences. First, a neutrino cannot be scattered by (say) a proton by electromagnetic means, since it has no charge, and therefore cannot emit or absorb a photon. However, it could well emit or absorb a Z^0, and could then undergo scattering. Moreover, charged particles, such as electrons and protons, could interact by exchanging Z^0s as well as photons. Although the contribution of the Z^0 exchange to the effective electron–proton interaction turns out to be fantastically tiny compared to that of photon exchange (about 10^{-16} in a hydrogen atom!), it has the parity violation property characteristic of the weak interaction, and hence should result in very small left-right asymmetries even at the atomic level. Both these effects have now been well confirmed experimentally, and provide strong circumstantial evidence for the correctness of the idea of electroweak unification. (Incidentally, it is worth remarking that these tiny asymmetries in atomic physics would almost certainly never have been looked for in the absence of the stimulus provided by the electroweak theory; and had they been accidentally observed, would almost certainly have been dismissed out of hand as due to some unknown experimental error.) To cut a long story short, the existence of the Ws and the Z^0, which had been believed by most people on theoretical grounds for some years, was finally established directly in accelerator experiments in 1983–4, when these particles were observed as resonances, roughly at the energies predicted, in proton–antiproton scattering, thus putting the final seal on the unified electroweak theory.

There is one very serious difficulty regarding this theory, however, and it is with attempts to resolve this difficulty that the fun really starts. The difficulty is this: if the $W^{\pm}$, and more particularly the Z^0, are really first cousins of the photon, how is it that the weak interaction is so short-range while the electromagnetic interaction is long-range? If we go back to the argument given earlier (p. 40) about interaction by exchange of virtual particles, we see that the electromagnetic interaction is long-range only because the photon is massless. Were it to have a mass, m, the energy needed to create it would be at least mc^2, so the time for which it would exist would be $\sim h/mc^2$, and the maximum distance over which it could propagate and give rise to an interaction would be $\sim h/mc$. Turning this argument around, we see that if an interaction of range r is to be propagated by exchange of virtual particles, those particles must have a mass which is proportional to $1/r$. From the very short range of the weak interaction we infer that the IVBs must be very massive—and their experimentally observed masses are in fact 81 GeV for the Ws and 93 GeV for the Z^0.

At first sight, one might just shrug one's shoulders and conclude that that's the way it is, that the IVBs 'just happen' to have large finite masses while the photon has zero mass. Unfortunately, apart from a certain lack of aesthetic appeal, this hypothesis runs into apparently insuperable technical difficulties. The resolution which most people currently accept looks, at first sight, very curious indeed. It consists of postulating a further boson field, the so-called Higgs field, corresponding to a so far unobserved particle, which interacts with the IVBs by a gauge mechanism, and which, moreover, undergoes the phenomenon, introduced earlier, known as spontaneously broken symmetry. That is, we assume that this field lies in some abstract space (in this case two-dimensional), and that the Lagrangian (and hence all interactions) is invariant under rotations in this space, just as the interactions in the idealized model of a ferromagnetic material introduced earlier are invariant under rotations in ordinary three-dimensional space. However, just as the magnetization in the ferromagnet develops at low temperatures a finite value which points in a given direction, so the Higgs field acquires a finite average value which points in a definite 'direction' in the abstract space. (Incidentally, just as in the case of the magnetic material, the effect is predicted to disappear at sufficiently high temperatures—so high that they are of interest only in a cosmological context.) One major difference with the case of the ferromagnet is that, as we saw, there is always in real life some small environmental field which 'chooses' the correct direction for the magnetization; whereas in the case of the Higgs field there is, at least in most current theories, no analogous effect. Fortunately, this does not matter (at least prima facie), since it turns out that the absolute 'direction' of the Higgs field is not observable and may not be physically meaningful—only changes of it in space and time have a direct physical meaning.

How does this help? Because if we start by assuming that our IVBs are massless, like the photon, and then allow them to interact with the Higgs field by a gauge mechanism, it turns out that they acquire finite mass. Moreover, the way in which they acquire it guarantees that the technical difficulties mentioned above go away. This mechanism, the Higgs mechanism for the generation of mass in otherwise massless particles by spontaneous symmetry breaking, is at the core of many current theories in particle physics; it is conceivable that *all* mass is generated in this way. (It is amusing that, historically, the phenomenon was first clearly recognized in the context of solid-state physics, in the theory of superconductivity, and was only subsequently applied to particle physics—a fact which should give pause to those who would dismiss some areas of physics as 'merely derivative'!) Thus, we have solved the puzzle of the qualitatively different behaviour, with distance, of the electromagnetic and weak interactions—but at a cost: we live in a world in which the original symmetry is spontaneously broken, but may nevertheless never know it.

Thus, as we come into the mid-eighties we have a picture of the fundamental constituents of matter as fermions (quarks and leptons) interacting with one another by gauge-invariant mechanisms, sometimes non-abelian, involving the exchange of virtual intermediate bosons (photons, gluons, Ws, and Z^0s, and perhaps gravitons), which start off massless but may in some cases acquire mass by the Higgs mechanism. What obvious problems does this picture suggest?

In the first place, the success of the last two decades in unifying the electromagnetic and weak interactions obviously induces us to see if we can go further, and include also the strong interaction (and maybe eventually gravity) in the unified scheme. A number of schemes for doing this, generally known as 'grand unified theories', have been developed in the last ten years or so. Most of them build in the idea that originally quarks and leptons are two varieties of a single class of 'fundamental fermions', and indeed can convert into one another by emission of a massive gauge boson (just as, in the electroweak theory, the electron and the neutrino are two varieties of the same entity, the lepton, and interconvert by emission of the $W^{\pm}$). One observation which may merely be an accident of the range of accelerator energies available in 1987 but which, if not, gives some support to quark–lepton unification is that at present there are exactly three known quark generations (u, d; s, c; t, b) and three lepton generations (e, ν_e; μ, ν_μ; τ, ν_τ). There are fairly compelling arguments that, if the general idea is right, the gauge bosons in question must be very heavy indeed (mass $>10^{14}$ GeV) and are therefore way beyond any possibility of direct observation in the foreseeable future. However, their existence can nevertheless have observable consequences. For example, the simplest such grand unified theory predicts that, owing to the possibility of exchange of such a superheavy boson, the proton is not stable but can decay into a neutral pion and a positron. The lifetime of the proton is, even on this view, almost unimaginably long—more than 10^{30} years, or roughly 10^{20} times the age of the universe. But since one can collect vast numbers of protons (a cubic metre of water contains roughly 10^{29}), it is not hopeless to try to detect proton decay processes experimentally. In fact, over the last five years or so a number of such experiments, using hundreds of tons of water in deep underground mines or caverns (to prevent spurious cosmic-ray events contaminating the data), have been in progress. Generally, in such experiments one would expect to see at most a few decay events per year, but so far even these have not been seen; and at the time of writing it seems that the evidence is against at least the simplest versions of grand unified theories. Whether more complicated versions will stand the test of experiment remains to be seen.

Another obvious problem concerns the Higgs mechanism which is supposed to give the IVB's mass. The Higgs field, like the photon field, should correspond to a particle, and while the non-zero average value which it is supposed to take in the vacuum may not be observable, fluctuations around it should still appear as particles. (In the magnetic analogy used above, these would correspond, roughly speaking, to fluctuations in the magnitude of the magnetization.) So far, this 'Higgs boson' has not been seen experimentally. Does it really exist? If so, where should we look for it? Unfortunately, the electroweak theory as currently formulated puts very few constraints on the mass of the Higgs boson: it could be anything from a value corresponding to an energy available in existing accelerators to one well beyond our reach in the foreseeable future. The search for this hypothetical particle, so fundamental to the current theory, is one major motivation for building accelerators of yet higher energy. Until it is found, there must remain some nagging doubt about whether the Higgs mechanism is really the *only* way to generate masses for the $W^{\pm}$ and the Z^0.

As we have already seen, the current picture divides the fundamental particles into two classes: fermions (quarks and leptons), which are (at the present level at least) originally massive, and bosons (photons, IVBs, and gluons), which carry the interactions between fermions and are originally massless, though some of them may acquire mass by the Higgs mechanism. (The Higgs particle itself is in a special class with respect to this classification.) The accepted electroweak theory and the grand unified theories currently under discussion attempt to unify the fermions with one another and the bosons with one another. A more ambitious goal, obviously, is to unify fermions with bosons—that is, crudely speaking, matter with interaction. Theories which attempt to do this go under the name of 'supersymmetry', or, if they are constructed to satisfy the requirements of general relativity (see Chapter 3), 'supergravity'. Although the idea of a supersymmetric theory is very enticing, a major problem with such theories is that they predict the existence of a boson for every known fermion, and vice versa. (Thus, for example, the photon has a fermion companion—the 'photino'—and the quark has a boson companion—the 'squark'.) Unfortunately, none of these companion particles have so far been observed. This may be because they are supermassive, or because they interact only very weakly with ordinary matter, or for some other reason; but until at least some of the predicted 'ghost' particles are observed, supersymmetric theories as such are bound to remain speculative.

Incidentally, supersymmetric theories are not alone in predicting the existence of currently unobserved particles, sometimes in very large numbers. One attempt to solve a technical problem regarding the strong interaction involves predicting the existence of a light boson called the 'axion', which may actually be all around us in much greater numbers than the photons in ordinary sunlight. However, because it interacts so extremely weakly with ordinary matter, we would almost certainly not have noticed this flood of axions unless we did an experiment specifically to look for it. A number of such experiments are now in progress. A rather similar status attaches to the so-called magnetic monopole which is predicted by most types of grand unified theory; there may be a detectable flux of such monopoles in cosmic radiation, but it needs a special type of experiment to find them. The current searches for axions and monopoles are unusual in the context of particle physics in not requiring ultra-high energy (and ultra-expensive!) accelerators: they can be done using essentially only 'everyday' energies in an ordinary university laboratory. The detection of axions or monopoles (or for that matter photinos, squarks, and the rest) could have important implications for cosmology (see Chapter 3).

Let us imagine, for the sake of argument, that grand unification is as successful in the next two decades as electroweak unification has been in the last two; and, moreover, that the Higgs boson is observed experimentally and is found to have all the properties predicted for it, thereby removing the last vestiges of doubt about the Higgs mechanism. Would particle physicists at that point be able to pack up and go home, their job completed? Contrary to the impression given in some of the more breathless popular writings on the subject, almost certainly not. If the successful unification scheme is anything like any of those currently envisaged, a whole host of questions would remain. We have already seen that such theories leave the relationship, if any, between fermions and bosons an open question. In addition, such theories, while they reduce the number of arbitrary parameters (particle masses, interaction strengths, and so on) somewhat, still require a considerable number of properties to be put in 'by hand'.

For example, why are there just three quark–lepton generations? (Or are there?) Why do the quarks and leptons have the masses they do? Why is the mass of the postulated superheavy 'grand unification' boson ($>10^{14}$ GeV) different by a factor $>10^{12}$ from that of the electroweak IVBs (the $W^{\pm}$ and Z^0)? Are quarks and leptons themselves in some sense composite? Furthermore, while we can *describe* the abstract symmetry properties of the various interactions quite adequately in quantum field theory, the *reason* for their presence or absence is quite obscure. Why, for example, should there be a symmetry corresponding to electric charge? And why should it be quantized in units (or simple fractions) of the experimentally observed electron charge? Why do the weak interactions violate parity conservation? And what is the origin of the CP violation in some weak interaction processes? For that matter, why does the universe we know possess one time and three space dimensions?

One possible approach to at least some of these problems is to hope that, if we examine the structure of Lagrangian field theory closely enough, and impose some very wide and general constraints on it, we may actually find eventually that there is only one possible theory consistent with all these constraints. In other words, it just could be that, simply to get mathematical consistency in the description, the world may *have* to have a certain number of space and time dimensions and *have* to have a Lagrangian with certain symmetries built in (and perhaps others absent). Incidentally, it is not essential that the number of space and time dimensions in the fundamental theory should be three and one respectively: it is perfectly possible to start with a larger number and get some of them to 'compactify' (curl up) on a scale so small that we would never observe them. In the last two or three years, a class of theories (so-called 'superstring' theories, in which the basic entities are 'strings', or lines, rather than the points which correspond to the traditional particle picture) which in effect have this character have become very popular. The hope is that the constraints imposed on such theories solely by the need for mathematical consistency are so strong that they essentially determine a single possible theory uniquely, and that by working out the consequences of the theory in detail, one might eventually be able to show that there *must* be particles with precisely the masses, interactions, and so on, of the known elementary particles:

in other words, that the world we live in is the *only possible one*. This programme is still in its infancy, and it is certainly far too early to judge its chances of success.

Even supposing such a programme were successful, however, there would still be a very fundamental group of questions hovering over the whole field of particle physics: namely, *why* should Nature seem to want to be described by Lagrangian quantum field theory, and therefore to submit herself to the stringent constraints imposed by it? Is the formalism universally valid in fact, even when it seems to give rise to severe paradoxes (see Chapter 5)? What is the *meaning* of the abstract internal symmetries which the consistency of the formalism appears to require? Or are these questions themselves 'meaningless'? I shall return to this topic in the final chapter.

3

The universe: its structure and evolution

We human beings inhabit a small planet, ninety million miles away from a rather undistinguished star, which in turn is a member of a galaxy itself in no way extraordinary. Until recently, our physical excursions were confined to the surface of this planet, giving us a 'range' of, let us say, about 10^7 metres. In the last couple of decades we have managed to reach that planet's main satellite, the moon, thereby extending our range of wandering by about another factor of thirty; and we can now make and dispatch space probes which can transmit information back to us from the outer reaches of the solar system, about 3×10^{12} metres away. Yet we know, or think we know, a great deal about the properties and history of the universe out as far as 10^{26} metres—more than 10^{13} times the range of even our latest probes. How is this possible? If we imagine a microbe confined to the surface of a microscopic speck of dust floating in the middle of St Paul's Cathedral, the microbe's problem in inferring the properties of the cathedral, or even the earth as a whole, would be trivial compared with ours.

We can only pretend to make any progress at all if we assume that the laws of physics, as we can discover them here on earth, apply to the universe as a whole. As one eminent cosmologist puts it:

> Modern scientific man has largely lost his sense of awe of the universe. He is confident that given sufficient intelligence, perseverance, time and money he can understand all there is beyond the stars. He believes that he sees here on earth and in its vicinity a fair exhibition of nature's laws and objects and that nothing new looms up there that cannot be explained, predicted, or extrapolated from knowledge gained down here. He believes he is now surveying a fair sample of the universe, if not in proportion to its size, which may be infinite, yet in proportion to its large-scale features. Little progress could be made in cosmology without this presumptuous attitude, and Nature herself seems to encourage it, as we shall see, with certain numerical coincidences that could hardly be accidental.[1]

It is impossible to stress too much the fact that, in applying physics to the universe as a whole, we very often have to extrapolate laws which have been tested firmly and directly over only a very small range of density, temperature, and so on to conditions which are different by many orders of magnitude. Indeed, in many cases we have to invoke ideas which have not been tested directly at all, but are themselves inferred from a complex series of indirect experiments. This may be thought to be extremely dangerous, and perhaps it is; but it is difficult to see how we could proceed otherwise, and at least we are in good company. Arguably, no single extrapolation in modern cosmology equals in intellectual daring the colossal leap made by Newton when he inferred the dynamics of the planets from the effects of gravity as observed on earth. It is worth reminding ourselves that until the advances in rocket technology of the last two or three decades this extrapolation had never been tested directly, in that we could not conduct controlled experiments but had to use what Nature deigned to give us. Yet the 'celestial mechanics' of Newton had already been dogma for two centuries, and very few people indeed seriously expected that Voyager II would fail to reach Uranus because, for example, new forces come into play on the scale of the solar system, which are not measurable (or have not been measured) here on earth. In the case of cosmology, direct and controlled tests will probably not be possible in the foreseeable future. With the current state of rocket technology, it would take a space probe about ten thousand years to reach even the nearest star other than the sun; and even if accelerated to almost the speed of light probes of even the nearest regions of our own galaxy would produce results available only to our remote descendants. Nevertheless, many contemporary cosmologists probably look on at least some of their ideas about the universe as a whole, rightly or wrongly, with the same certitude as mid-nineteenth-century astronomers did on the Newtonian theory of the solar system. It might therefore be worthwhile to remind ourselves that a tiny discrepancy between Newtonian theory and observation—on the precession of the perihelion of Mercury, see below—which was long dismissed as an irritating but minor anomaly, now, with the benefit of hindsight, signals that the whole edifice was unsound—or, at least, was not the whole story. How many similar tell-tale discrepancies are there in today's cosmology, which our descendants will say should have told us that our physics was incomplete? We don't know, of course; and the only practical way of proceeding is to carry on applying physics as we know it here on earth until experiment comes up with a really spectacular contradiction to our predictions.

Given, then, that we propose to use our terrestrial physics to interpret the cosmos, what kind of information can we get, and how? As already pointed out, we have to rely on what Nature gives us. Apart from a few samples of bulk matter of relatively local origin (meteorites, dust from the moon, and so on), what Nature gives us are simply the particles and electromagnetic waves arriving in the earth's vicinity from outer space. Not all this material reaches us at the earth's surface: for example, X-rays are strongly absorbed by the atmosphere, which is why the ability we have developed in the last two decades or so to send satellites equipped with appropriate detection equipment into orbit has resulted in a considerable increase in our information.

Among the massive particles arriving from outer space, the majority are protons, though with some admixture of electrons and the nuclei of helium and heavier elements. The information provided by these particles tends to be rather difficult to interpret. Being massive, they do not travel at the speed of light, or necessarily in straight lines (for example, they can be deflected by the galactic magnetic field), so it is not easy to infer their origin or their history. Neutrinos do not suffer either of these drawbacks, but as we saw in the last chapter, they are very difficult to detect. As a result, most of our knowledge of the universe in the large comes from the study of the electromagnetic radiation arriving from outer space. In the early days of astronomy, the only part of the electromagnetic spectrum which could be used was the very small segment which is visible to the human eye; nowadays, with the help of satellites, long-baseline interferometers, and other specialized detection equipment, we can study the whole spectrum, from radio frequencies up to hard γ-rays (see Chapter 1). To see what kind of information we can extract, let's focus on the visible radiation (light) and suppose we are looking at a visible star (such as our own sun).

The light we can produce on earth (from candles, electric filaments, arc lamps, and so on) is emitted by atoms or molecules making transitions between their various possible states, and any given atom or molecule will, if suitably isolated, produce a characteristic set of spectral lines; that is, it will emit light only with certain special frequencies and the corresponding wavelengths. So, when we observe one of these patterns of lines in the light emitted by a star, it seems natural to infer the presence in the star of the element which we know emits such a spectrum on earth.[2] In some favourable cases we can infer not just the element, but the particular isotope of the element which is emitting the light; for example, it is possible to distinguish heavy hydrogen (deuterium) from the ordinary variety. In this way we can build up a table of the abundances of the elements in particular stars and also in interstellar and intergalactic space. Also, when a source produces light on earth, the intensity of the light emitted depends on the temperature of the source, and its polarization depends on the electric and magnetic fields present; so, extrapolating once more, we can calculate the relevant values of these quantities for the star in question. If we know the distance of the star and are able to measure the total power received, we can estimate its total power output. Finally, in some cases an interesting variation in the pattern of reception of radiation with time leads us to surmise that the source is performing some kind of regular or irregular motion; this is the case, for example, with the so-called pulsars, which are generally believed to be rotating neutron stars.

Putting all this information together, and using also the knowledge of nuclear physics acquired in the laboratory, we arrive at a picture of a typical star (such as the sun) as a giant nuclear power station, which generates its heat by the so-called nuclear fusion process in which hydrogen is converted by various steps into helium and heavier elements with a large release of energy (a process which, unfortunately, we have so far been unable to achieve in a controlled way, as distinct from an uncontrolled way in a bomb, here on earth). Although the surface temperature of such a star is only a few thousand degrees, the temperature of the interior, where the nuclear burning process actually takes place, is of the order of a few million degrees. At such temperatures ordinary matter as we know it in everyday life does not exist; all atoms have been completely ionized—that is, the electrons have been stripped away from the nuclei—and the density is about a hundred times that of terrestrial solid matter. This situation, which can be achieved on earth for a tiny fraction of a second in a nuclear bomb, is the permanent state of the interior of a typical star as long as its supply of nuclear fuel lasts. After the supply of hydrogen has all been converted into helium, the helium itself can be further converted into heavier elements, mostly carbon or iron, but at this point the process stops; no further energy can be extracted from the nuclear fuel, and the star cools and collapses inwards on itself, sometimes with spectacular death-throes. (As this book goes to press, just such a death-throe, a so-called 'supernova' explosion, is occurring, for the first time in 300 years, in a star in our own galaxy, and is visible with the naked eye from southern latitudes.)

Depending on the mass of the star, its final state may be as a so-called white dwarf, a neutron star, or perhaps a black hole. White dwarfs contain the constituents of ordinary terrestrial matter (nuclei and electrons) but packed to a density about a million (10^6) times that of typical liquids or solids on earth. Neutron stars, as their name implies, consist mainly of neutrons, but with some protons and electrons, and are packed to the incredibly high density of about 10^{15} times that of ordinary matter; this density is about three times that of the protons and neutrons in the atomic nucleus, and indeed, a neutron star can in some sense be regarded as a giant nucleus, whose radius is not 10^{-13} centimetres, but more like 10 kilometres! Clearly, both for white dwarfs and, even more, for neutron stars our laboratory experience of the behaviour of matter is quite inapplicable, and we have to start again more or less from scratch to work out the properties of extended matter at these enormous densities. In doing so, we make the usual conservative assumption that the basic laws of physics as discovered on earth will continue to hold—that is, that the mere compression of matter to high densities over large scales will not introduce qualitatively new basic laws. Since we cannot do controlled experiments on neutron stars, the plausibility of this assumption has to be tested by comparing the results of our calculations with the evidence that Nature provides in the form of the properties of the radiation emitted by them. So far, the theory has stood the test reasonably well; thus one can say that, while the observations do not exclude—and probably never could exclude—the possibility that new laws may be coming into play, they at any rate do not require it. There is, however, one proviso to be made here. It is generally believed that when the mass of a system divided by its radius reaches a certain critical value, there is generated the peculiar warping of space and time known as a 'black hole', and that at this point the laws of physics as we know them fail, or, more precisely, fail to provide any further useful information (I will return to this later). It turns out that if a star with a mass only slightly greater than that of our own sun were to collapse into the neutron-star state, the critical value would be reached, so that our familiar physics would be of no further use to us in this context.

By day, we humans are mostly conscious only of our own star, the sun; by night, we can make out with the naked eye perhaps a few thousand stars, of which it turns out that the nearest (other than the sun) is already about 4 light-years, or 4×10^{13} kilometres away from us. (I discuss below how we know this.) With the aid of modern telescopes, we now know that the visible stars are but a tiny part of a group of about 10^{11} stars—the Galaxy—which are distributed individually in a fairly random way, but so that their density is highest in the plane of the Milky Way (hence the name, *gala* being the Greek word for 'milk'). The diameter of the Galaxy is very roughly 10^{21} metres (10^5 light-years), and our own sun lies about two-thirds of the way from the centre to the (diffuse) outer edge, roughly in the plane of maximum density. (This is, of course, why we see the Milky Way as a region of increased density of stars.) For many years, it was widely believed that our galaxy was in effect the whole universe; but we now know that it is itself only one of a cluster of galaxies, the so-called local group (the name is somewhat ironic, since our best-known neighbour, the Andromeda galaxy (nebula) is about two million light-years away!); and that the local group is itself part of a 'supercluster', which is again only one of innumerable superclusters which fill the universe as far out as our telescopes can reach. We appear to be in no way in a privileged position.

In fact, if we look at that part of the contents of the universe which is concentrated in stars, there are two properties which at first sight seem rather trivial, but are actually very remarkable. In the first place, we find structure on a local scale: matter is concentrated into stars, stars are clustered into galaxies, galaxies grouped into clusters, and these again into superclusters. There are also huge regions ('voids') where the density of galaxies is very much less than the average. The study of how these large-scale structures might have formed is at present a very active area of cosmology. On the other hand, on the largest scale accessible to our observations the universe appears *on average* to be both homogeneous and isotropic: that is, on average, all regions of the universe, whether close to us or distant from us, appear to be much the same in their properties (homogeneity), and no particular direction is in any way singled out (isotropy). Note that homogeneity does not imply isotropy; for example, it could have turned out that galaxies were distributed at random over all of space, but that they tended to have their planes of maximum density predominantly (say) parallel to that of the Milky Way. This seems not to be the case. The universe as we know it is both homogeneous and isotropic. As we will see, these properties persist when we take account of the matter which is not concentrated in stars: they are a crucial feature which any satisfactory cosmological theory has to explain.

As a matter of fact, it is not at all clear that all, or even most, of the matter in the universe is actually concentrated in stars. Certainly, within our own galaxy (and hence presumably within other similar galaxies) there is a considerable amount of 'interstellar dust' (material in the space between the stars); most of this is in the form of atoms or molecules of hydrogen, but there are also heavier atoms and molecules (including, fascinatingly, some of the amino acids crucial to the evolution of biological systems). These atoms and molecules can be detected by their effect in *absorbing* electromagnetic radiation of certain characteristic wavelengths. Whether or not there is any appreciable amount of matter in intergalactic space (that is, space between galaxies or between galactic clusters, as opposed to space between the stars within a galaxy) beyond the amount (about 1 atom per cubic metre!) already observable by its effect on light absorption, is still an open question. Quite independently of this question, however, a detailed study of the stability and motion of galaxies as a whole points strongly to a remarkable conclusion: much of the mass of galaxies exists in a form which has so far not been detected! This is the so-called 'missing mass', or 'dark matter' problem. Like most problems in cosmology, it can be avoided, at a cost—for example, by postulating new forces, or a modification of old ones, on the galactic scale—but at present most cosmologists take it fairly seriously. A great variety of candidates for the 'dark matter' have been proposed: some of them are types of elementary particles already postulated in a particle-physics context (massive neutrinos, axions, photinos, and so on); others come out naturally from the general theory of relativity (black holes); and no doubt in future others as yet undreamed of will appear. At the moment, however, we quite literally do not know what (most of) the universe is made of.

So far we have acted as if all the electromagnetic radiation we receive on earth has been emitted by stars, or perhaps by interstellar matter. But this is not actually so. One of the most remarkable advances of the last twenty-five years in cosmology has been the discovery of the so-called cosmic blackbody radiation. Blackbody radiation refers to the flux of electromagnetic radiation which one would get in thermal equilibrium at a constant temperature (see Chapter 1); it has a very characteristic pattern of intensity as a function of frequency (or wavelength), and from the value of the wavelength at which the intensity is a maximum, one can read off the temperature. It was discovered in 1965 that the earth is bathed in a flux of electromagnetic radiation whose intensity distribution is to a high degree of accuracy that which would be characteristic of a temperature of about 3 degrees absolute; the wavelength of maximum intensity lies in the microwave region of the spectrum, at about 1 millimetre. It is noteworthy that, once one allows for the motion of the sun relative to nearby stars and so on, the pattern of radiation seems to be isotropic within the accuracy of current measurements—that is, the same amount arrives from all directions. The existence, magnitude, and effective temperature of this cosmic blackbody radiation is another crucial datum of cosmology.

As we saw in the last chapter, in quantum theory, electromagnetic radiation is carried in the form of 'chunks', called photons. From a measurement of the cosmic blackbody radiation, one can obtain directly the number of photons per unit volume associated with it in our neighbourhood. Within our galaxy, light emitted by the stars also contributes photons, but on the scale of the universe as a whole this effect should be negligible. Therefore, if we assume that the cosmic blackbody radiation observed here on earth is characteristic of that in intergalactic space (and at present we know of no convincing reason not to assume this), we can estimate the number of photons per unit volume in the universe as a whole. A useful quantity to compare this with is the number of baryons per unit volume, again averaged over the observable universe, this comparison having the advantage that there are theoretical reasons for believing that neither number has changed very much in the recent past. It turns out that the ratio of photons to (observed) baryons is about 10^9 to 1. (The baryons are predominantly protons, either isolated or in the form of atomic hydrogen.) This is another observation which any history of the universe has to try to explain.

Let us note one more remarkable fact: namely, that the universe as we know it seems to contain very little antimatter (see Chapter 2). To be sure, this is a matter of not totally rigorous inference: it is practically impossible to distinguish antimatter directly from matter (by observing the properties of the radiation it emits, for example), and one infers the absence of antimatter primarily from the lack of evidence, in the form of copious γ-rays, for processes of matter–antimatter annihilation (and the absence of appreciable quantities incident on earth from outer space). This argument might perhaps fail if the 'antimatter galaxies' were sufficiently separated from ordinary ones. If there is indeed very little antimatter in the observable universe, this poses a considerable problem, since at the particle-physics level, matter and antimatter enter on a (nearly) completely equal footing. (We will see, in Chapter 4, that a rather similar situation arises with regard to the chirality of biological molecules.)

From what I have said so far, the reader might get the impression that most of the interesting activity in the universe, and in particular, most of the energy production, goes on in stars. Until a few years ago, this would have been the conventional view. One of the most exciting discoveries in astronomy over the last couple of decades, however, has been that many galaxies have active regions near their centres where massive amounts of energy are produced in processes which appear to be quite different from the ordinary nuclear reactions which we believe to occur in stars. There is no general agreement at present about the mechanism of this energy production, but one plausible view is that it is a result of matter falling into a massive black hole (see below).

We still have not mentioned what to most cosmologists is probably the most important single fact about the universe—

namely, that it is expanding. Since the great majority of current theories about the genesis and possible future of the universe rest crucially on this belief, it is as well to spend a few moments considering the evidence for it. We first consider the question: How do we know how far away astronomical objects are? In ordinary life on earth we have a number of ways of judging distance (other than by measuring it directly with a ruler). One way is to use (usually, of course, quite unconsciously) the fact that our eyes are a few inches apart; this is useful only for objects within a few feet of us at most. A second way is to 'send a probe', that is, project towards the object in question a small object whose motion in time is known, and measure how long it takes to arrive; for example, we can drop a stone down a well and count the seconds until the splash. A third possibility is to use 'parallax'—that is, to move our position and note how the object in question appears to move against the distant background. Finally, we very often judge the distance to far-away objects simply from a knowledge of how large they are. With a view to the astronomical analogies, we should note the rather trivial fact that for this last method to work even approximately, it is essential to know what kind of object we are looking at (a man, a tree, a house, or whatever); any mountain hiker knows how easy it is, in mist, to mistake a nearby rock for a distant mountain, or vice versa.

All these commonplace methods of judging distance have astronomical analogues, although some of them are useful only in a very restricted area. The analogue of relying on binocular vision is to take theodolite readings of the apparent position of the object in question simultaneously from different points on the earth's surface and use triangulation; this works only for very close astronomical objects such as the moon. The analogue of dropping a stone down a well is to send an electromagnetic (radar or laser) pulse towards the object in question and time the reception of the reflected pulse; since we know the speed of light, this gives us the distance directly. This method can be used for most of the objects in the solar system; beyond this, it is not practicable, since, quite apart from the fact that a radar wave reflected from the nearest star would take years to return, the reflected wave would be so weak as to be undetectable. The possibility of a parallax method is provided by the motion of the earth around the sun; by measuring the apparent position, relative to the so-called fixed stars, of the closest stars at a time interval of six months or so, we can infer their distances from us. Given current limits on the accuracy of direction determination, this method works reasonably well for objects up to about 300 light-years away (at which distance the earth's orbit looks, roughly, like a penny viewed from a distance of sixty miles!). In principle the range of this method can be extended by taking observations from space probes sent out to the far reaches of the solar system, but the gain is not great.

A range of 300 light-years, alas, does not even cover most of our own galaxy, let alone anything beyond it. Beyond this range, we have to resort to more indirect methods, which require us in some sense to know what we are looking at. There are a number of such methods, and it is the fact that, by and large, they give distance estimates which are mutually consistent, rather than the intrinsic reliability of any one method by itself, which gives us confidence in the results. As an illustration of methods of this type, let us consider the one based on the properties of a class of stars known as Cepheid variables (from the best-known member of the class, δ Cephei). There are a number of such stars which are close enough to us that we can establish their distances by the parallax method; hence, we can infer from their apparent brightness their absolute brightness (luminosity)—just as by measuring the light received from a light bulb at a known distance, we can work out its power. Now, Cepheids are remarkable in that they show a well-defined pattern of variation of brightness with time. Moreover, a study of those which are close enough that we can estimate their absolute luminosities reveals a close and rather accurate relationship between the (average) absolute luminosity and the period of the oscillation—that is, the length of the cycle in the brightness variation. The astrophysical reasons for this relationship are well understood, and it therefore seems reasonable to assume that it holds for *all* Cepheids, whether close to us or not. If so, then by observing the period of a distant star of this class, we can work out its absolute luminosity, and then, by comparing this with its observed brightness on earth, infer its distance from us (just as, if we know that a certain light bulb has a power of 100 watts, we can work out its distance from us from its apparent brightness). In this way the distance of Cepheid variables, both within and outside our own galaxy, can be determined out to a distance of about three million light-years. Of course, most stars are not members of this special class, so to use this method to determine the distance from us of a far-away galaxy, we have to assume that any Cepheids which lie in the direction of that galaxy are actually part of it and do not just happen to lie in that part of the sky by accident. Since the statistical probability of such a random coincidence is extremely small, this seems an eminently reasonable assumption (just as it seems very reasonable to assume that when (say) a strong radio source lies in a direction indistinguishable from that of an optically observed galaxy, the two are in fact the same object, or at any rate closely associated). Here, as so often, the cosmologist puts his faith in Einstein's dictum that 'God is subtle, but He is not wickedly perverse'. As remarked above, this conclusion is strengthened by the fact that other independent arguments of the same general type often lead to similar estimates of the distance of far-away galaxies. Beyond a few million light-years, where this and similar methods fail, the determination of cosmic distances is even more indirect, as we will see below.

So, we suppose that we can reliably infer the distance from us of remote objects at least up to this limit. What of their velocities relative to us? As we shall see, the presumed velocities of most astronomical objects correspond to a fractional change in their distances from us of about 1 part in 10^{10} per year. This is far too small to be detectable by any of the above methods (except conceivably in the case of the planets, where, of course, the cosmological effects to be discussed are totally dwarfed by the obvious gravitational effects on the motion). In fact, the velocities of most astronomical objects relative to us are inferred by a quite different method, which relies on the so-called Doppler effect: namely, if a moving object emits a wave, be it a sound wave, a light wave, or any other kind, then the frequency we detect is lower or higher than that of the wave as emitted, according as to whether the object is moving away from us or towards us. A well-known example of this effect is the way in which the pitch of an ambulance siren seems to drop abruptly as the ambulance passes us. The reason is fairly easy to see: if the siren emits 300 wave crests per second, and the ambulance is travelling away from us, let us say, at 100 feet per second, then the last wave crest of a given second is emitted about 100 feet further away from us than the first, and since the speed of sound is about 1,000 feet per second, it takes one-tenth of a second longer to reach us. So we actually hear the 300 wave crests arrive not in one second, but in 1.1 seconds, and so judge the frequency of the sound to be only 300/1.1, or about 270 cycles per second; in other words, we hear a pitch that is lower than the one emitted. To be sure, this simple argument works only for those kinds of waves, such as sound waves, which actually represent a disturbance in a physical medium (in this case, air). The argument concerning light waves is more sophisticated, and has to take account of the modifications of our concepts of space and time introduced by both special relativity and (when applied to make inferences on a cosmic scale) general relativity (see below), but the upshot is the same: namely, if we know the frequency (or wavelength) with which a light wave was emitted, and we measure the wavelength of the light we observe, then we can infer the velocity of the emitting source relative to us (or, more accurately, the component of the velocity which is directed towards us or away from us). This conclusion has been tested innumerable times in terrestrial laboratories; as usual, we are going to extrapolate it boldly into outer space.

We are very familiar with the fact that here on earth each element (more precisely, each element in a specific state of ionization) emits visible light with only certain specific wavelengths; the pattern of wavelengths acts as a sort of fingerprint of the element in question. Now, if we look at the light emitted by a reasonably close galaxy, we find that we can identify the same patterns which we recognize as characteristic of various elements here on earth, but that there is a small shift in the wavelengths; generally, the wavelength observed in the light from the galaxy is somewhat longer than that observed on earth, and the spectral pattern is therefore said to be 'red-shifted' (red light corresponding to the longest wavelength in the visible spectrum). The most natural interpretation is that these galaxies are moving away from us. As we go to more and more distant galaxies (inferring their distance in the way explained above), the 'red-shift' increases, and a quantitative investigation yields a quite remarkable conclusion: namely, that the galaxies appear to be receding from us at a speed which is directly proportional to their distance from us. This is the famous law announced by

Edwin P. Hubble in 1929, and bearing his name. The ratio of distance to velocity of recession is known as the 'Hubble time'; clearly, if the speed of recession were constant with respect to time, it would be just the time which has elapsed since all galaxies were collapsed together in our immediate neighbourhood. The currently accepted value of the Hubble time is about 1.5×10^{10} years, with an uncertainty of about 30 per cent in each direction; we will return below to the significance of this number.

The so-called cosmological red-shift described above and its interpretation in terms of the recession of galaxies from us—that is, in terms of the expansion of the universe—is such a fundamental datum of modern cosmology that it is worth stepping back for a moment to ask how firmly it is established. Clearly, the quantitative interpretation relies heavily on the correctness of the estimates of cosmic distances; while it is conceivable that they are in serious and systematic error, the amount of indirect but interlocking evidence on this point is sufficient that few people at present worry seriously about it. A more delicate question is the possibility of alternative interpretations of the observed red-shift. In fact, at least one other cause of red-shifting of light is known: namely, the gravitational effect postulated by the general theory of relativity (see below); and in principle, one might suppose that the red-shifted galaxies are sitting in a very strong gravitational field, while almost stationary with respect to us. There are quite strong arguments against this hypothesis, the principal one being that if this were the case, and if we are not in a specially privileged position in the universe, then the distribution of red-shifts between the observed sources should be different from what is observed. A more difficult hypothesis to refute is that the laws of physics themselves may change systematically with (for example) position in the universe, in such a way that the 'same' atomic transitions produce spectral lines whose frequency depends on the cosmological conditions in the region of the emitter; in that case, the red-shift would be no evidence for motion away from us. At present, all we can say about this hypothesis is that it does not appear to be forced on us by the experimental evidence (at least not yet!), and that the principle of Occam's razor recommends that we try to get along without it as long as possible. Finally, we shall see that even our currently accepted ideas about space and time on a cosmological scale actually give these concepts a meaning very different from the everyday interpretation adequate for going about our terrestrial business, and that it can by no means be excluded that future theoretical developments will give them yet another twist, so that, for example, it might even turn out that the cosmologists of the twenty-second century will regard the question 'Is the universe expanding?' as just as meaningless as we regard questions about absolute simultaneity. All that can be said at present is that the vast majority of cosmologists do regard the question as meaningful, and most (though not all) hold that the evidence in favour of the answer yes is overwhelming. Indeed, such is the general confidence in this interpretation of the data that in the case of very distant objects whose distance from us cannot be measured in any other way, it is now almost standard to *infer* the distance from the measured red-shift, and to use the inferred distance together with the observed luminosity to determine the absolute luminosity.

Before going on to discuss the modifications of our ideas about space and time which we believe to be necessary if we are to describe the cosmos, let us take a moment to ask: Supposing we were just to continue to operate with our classical (Newtonian) notions, how would we describe the universe in the light of the evidence for its expansion? First, could we conclude that, since almost all galaxies appear to be receding from us, we must be in the centre? No. If a rubber sheet is uniformly stretched, an observer at any point on the sheet will see all neighbouring points receding from him, with a velocity proportional to the distance. Second, what could we conclude about the history of the universe? To conclude anything at all, we would need to know something about the dynamics—that is, the forces acting on the system and its reaction to them. The simplest assumption is that, on the cosmic scale, as well as the scale of the solar system and that of the falling apple, the only important force affecting matter in bulk is gravity, and that not only Newton's law of universal gravitation, but also his laws of motion apply. If we also assume that on a sufficiently large scale the distribution of matter in the universe is homogeneous, and ignore some nasty conceptual difficulties associated with what goes on at the edges (the so-called pseudo-Newtonian model), then the idealized problem so defined can be completely worked out, and the conclusion is that the universe did indeed have a 'beginning'—that is, there was a time in the past at which all galaxies were packed together with essentially infinite density. Or, more precisely, there was a time before which we cannot expect our approximations to apply and don't really know what happened (see below). The 'age of the universe' determined in this way is of the same order as the Hubble time, though not in general actually equal to it: that is, it is about ten billion years. This is rather interesting for the following reason. Certain heavy elements found in the earth's crust, such as uranium, are believed to have been produced about the time the earth was formed, and thereafter to have decayed radioactively. From a study of the present abundance of those elements and the elements produced by their decay, we can reasonably infer that the earth must be at least about five billion years old (and is probably not much older). A similarly circumstantial argument based on an analysis of the energy-generating process in stars leads to the conclusion that most stars are very roughly ten billion years old. The apparent consistency between these figures increases our confidence in our simple estimate of the age of the universe.

What of the future? Will the universe continue to expand for ever? It turns out that within the pseudo-Newtonian model there are three possibilities. First, the velocity of expansion, while decreasing with time, may always remain above a certain finite positive value, so that however far we go into the future the distant galaxies will recede at a rate at least equal to this. Second, the rate of expansion, while remaining positive, may become smaller, and eventually tend to zero, so that in the distant future we approach ever more closely a static situation. Third, the universe may reach a maximum size (or rather, expansion; see below) and then begin to contract again, ending up in a state of infinite density similar to its origin ten billion years ago. (What happens after that—if there is an 'after'—is quite literally anyone's guess.) Which of these three scenarios is realized depends critically on the present density of the universe as a whole. If the density is exactly equal to the so-called critical, or closure, density—about 2×10^{-29} grams per cubic centimetre—the second scenario is realized; if it is less, the first; and if it is more, the third. A catalogue of the currently observable constituents of the universe (see above) yields a mass density which is about 1 to 2 per cent of the closure value; if this is indeed the true mass density, then the universe is destined to expand for ever. However, there are quite strong theoretical reasons for believing that we are actually missing a quite substantial fraction of the mass (the 'dark matter' mentioned above), and that the true mass density is at least 10 per cent of the closure value, and could perhaps even be equal to it. Here, unfortunately, as so often in cosmology, uncertainty in one experimental number affects the whole edifice. The critical density varies as the inverse square of the Hubble time, and is therefore itself uncertain to within a factor of about four. So we just do not know at present whether the density is below, above, or equal to the closure value.

One other obvious question which must have occurred to the reader is: Is the universe infinite in extent, or does it have an edge, beyond which there is, quite literally, nothing? The reason I have not mentioned this matter so far is that the very nature of the question is changed rather radically by the conception of space and time on the cosmic scale which follows from the special and general theories of relativity, to which I now turn.

To Newton, the first person to investigate quantitatively the mechanics of extraterrestrial bodies, both space and time were absolute concepts: there existed a unique frame of reference for the measurement of position in space—namely, the frame of the fixed stars—and also a unique scale of time. Of course, just where one sets the origin of the spatial reference frame and what time one chooses to call zero were arbitrary. But apart from that, there was no freedom: the distance between points in space and the interval between points in time were uniquely defined and unambiguous concepts. Actually, well before Newton, Galileo had appreciated, in a terrestrial context, the relative nature of the space frame of reference. He had observed, for example, that on a steadily moving ship (or, as we would probably prefer to say nowadays, an aircraft), all mechanical phenomena would look the same as in a stationary frame. (In an accelerating frame of reference, however, things genuinely look different; we can all tell when a plane is braking or turning.) So the space frame is to that extent arbitrary; and in particular, the distance between two events happening at different times will depend on the observer. Successive flashes by an aircraft's wing lights at night look to a passenger on the plane as if they occur at the same point, but an observer on the ground sees them at different points. But

Galileo, Newton, and their successors for two hundred years did not doubt that *time* intervals were uniquely defined, and were independent of the observer: the observer on the ground, once he has allowed for the time taken by the light to reach him, should infer the same time interval between the flashes as the passenger in the aircraft.

As the theory of electromagnetism was developed during the nineteenth century, however, and as it was appreciated that light was an electromagnetic wave, it became clearer and clearer that there were great difficulties in reconciling these concepts with what was known about light propagation. In particular, almost all the experimental observations seemed consistent with the hypothesis that electromagnetic phenomena, as well as mechanical ones, looked the same in any frame of reference, and in particular that the velocity of light looked the same, however the observer was moving (uniformly) relative to the fixed stars. But this observation is very difficult to reconcile with the absolute nature of time. Consider, for example, a flash emitted by a light at the nose of an aircraft and received at the tail. Because of the finite speed of light, in the time between emission and reception the aircraft will have moved a small distance, and therefore the spatial separation between the two events will be perceived to be different by a passenger and a ground-based observer. On the other hand, they are supposed to agree precisely about the time interval. If so, however, they will draw different conclusions about the speed of light—in apparent contradiction to experiment. Thus, if the speed of light is indeed the same for all (uniformly moving) observers, they cannot measure the same time intervals between the two events.

(At the risk of stating the obvious, I might emphasize that I can only make this brisk argument, and have a reasonable expectation that the reader will accept it, because we have both had the advantage of eighty years' worth of (conscious or unconscious) indoctrination in relativity theory. To physicists in the late nineteenth century and the first decade of the twentieth, the conclusion did not seem obvious at all; and indeed they went through what from a modern vantage-point look like absurdly contrived intellectual contortions to avoid it. Will some of the current fashions in cosmology and particle physics look equally contrived a hundred years from now? Probably.)

As everyone knows, Einstein cut the Gordian knot in his special theory of relativity in 1905. This theory concerns the relationship between physical phenomena as viewed by observers moving with uniform velocity relative to one another. In essence it simply says: yes, the speed of light *does* look the same to all such observers; and no, they do *not* in general measure the same time interval between events, Δt, any more than they measure the same spatial interval, Δx. In fact, time and space intervals are intimately related, and there is a particular combination of them which is indeed the same for all observers: namely

$$(\Delta x)^2 - c^2(\Delta t)^2$$

where c is the speed of light, and is a universal constant (3×10^8 metres per second). (It follows from this that if Δx is larger (smaller) than $c\Delta t$ for any one observer, it is so for all observers.)

The special theory of relativity has many well-known implications, some of which have already been referred to in Chapter 1. Among them are: that mass and energy are interchangeable; that it is impossible to accelerate a body with finite mass to the speed of light; that no causal effect can propagate faster than light, so that two events which occur with spatial separation Δx and time separation Δt such that $\Delta x > c\Delta t$ cannot influence one another (such pairs of events are said to have 'spacelike separation'); and that processes taking place in a system moving with high velocity with respect to the earth (for example, decay processes in a cosmic-ray muon) appear, as observed on earth, to 'go slow'. All these predictions (which are of course made quantitative in the theory) have been repeatedly tested and confirmed in laboratory experiments, sometimes in quite spectacular ways. For example, with modern atomic clocks it has been possible to verify experimentally the famous 'Twin Paradox'—that a person (or at least a clock) who travels and returns to his base ages less than one who stays at home. It is probably no exaggeration to say that special relativity is the most firmly grounded component of the whole of the modern world-view of physics, and probably the single element which most physicists would be most reluctant to give up (see Chapter 5).

The situation is rather different when we come to the *general* theory of relativity, formulated, also by Einstein, in 1916. In

Newton's mechanics, the concept of the mass of an object plays two quite different roles: it determines the acceleration of the body when acted on by a given external force ('force equals mass times acceleration'), and it gives the force exerted on the object by gravity. It is convenient to distinguish these two aspects of the concept of mass by the terms 'inertial mass' and 'gravitational mass' respectively. That inertial mass is in all known cases equal, or, more accurately, proportional, to gravitational mass was already known in effect to Galileo; it is the point of the (almost certainly apocryphal) story of his dropping lead cannon-balls of different sizes simultaneously from the Tower of Pisa, since the proportionality in question leads to the conclusion that in the absence of frictional effects and so on the balls would indeed accelerate at the same rate, and therefore hit the ground simultaneously.

Over the succeeding three centuries the equality of inertial and gravitational mass was never seriously questioned; but it took Einstein to make this everyday fact the foundation of a new way of looking at space and time. Einstein observed that, since all bodies are accelerated at the same rate, a person falling freely in a uniform gravitational field would have no way of telling, *merely by doing experiments in mechanics*, that he was not floating in free space under the action of no force at all. (Fifty years later, this phenomenon was put to practical use in acclimatizing astronauts training for moon flight to the 'gravity-free' environment by putting them in planes in free fall in the earth's gravitational field. This actually exploits not the principle itself, but a corollary of it: namely, that in a frame falling freely under a gravitational force, phenomena are independent of how strong or weak that force is; on the way to the moon one is not actually gravity-free, but may be falling freely in much reduced gravity.) Then, by the kind of conceptual leap so characteristic of his thinking, he postulated that such an experimenter would have *no way at all* of telling that he was not in force-free space. This is the so-called equivalence principle. Since the methods at his disposal would certainly include optical ones, this principle immediately implies that the path of light must be 'bent' by a gravitational field. Consider a person doing optical experiments in a freely falling lift. Since he is supposed to be unable to tell that he is not in force-free space, and since in free space light certainly travels in straight lines, he should also see the light rays travelling in straight lines; for example, if a well-collimated beam passes through three successive equally spaced apertures, then according to him the positions of the apertures must lie on a single straight line, which for definiteness we suppose to be horizontal. But the light takes a finite time to travel between the apertures, and since the lift is moving relative to the earth, an observer fixed to the earth will record that the three apertures were at different vertical positions relative to him when the light passed through each of them. Moreover, the lift is not moving uniformly, but is accelerating, so that the vertical distance he measures between the second and third apertures at the time the light passed them will be greater than that between the first and second. Thus, the path he observes for the light cannot be a straight line: in fact, by adding more and more apertures (so as to trace the path continuously) and doing a little simple geometry, it becomes clear that the path must be the arc of a circle of radius c^2/g, where c is the speed of light and g the local acceleration due to gravity. In effect, the light ray is also 'falling' in the gravitational field, just like an ordinary particle! Since c is about 3×10^8 metres per second, and g at the earth's surface is about 10 metres per second per second, the 'radius of curvature' of the path (the radius of the circle of which it is a segment: not to be confused with the local curvature of space, on which see below) is about 10^{16} metres. At the surface of the sun, where gravitational acceleration should be about thirty times that on the earth, the corresponding radius of curvature should be about 3×10^{14} metres. Since this is very much larger than the sun's radius, and the acceleration falls off fast with distance from the sun, the net effect is that a light ray from a distant star approaching the sun at grazing incidence should be very slightly bent (by about 2 seconds of arc, it turns out). The experimental observation of this effect (in the solar eclipse of 1919) was one of the first and most spectacular confirmations of the general theory of relativity.

A second fairly direct consequence of the principle of equivalence is the gravitational red-shift. Imagine a lift released from rest in a gravitational field (say the earth's) which provides an acceleration g. A light flash is emitted from the roof of the lift at the moment of its release, and is then detected at the floor. Suppose the spatial distance between the events of emission and detection is d. Then the time taken by the light to travel this distance is d/c; in the meantime the lift has acquired a velocity, v, downwards equal to gd/c. But, as we already saw, this means that an observer falling with the lift will see a Doppler shift; in fact, if v is small, then according to the argument given above (which is adequate for our present purposes), the apparent frequency he sees will be shifted downwards by a fraction v/c. On the other hand, the principle of equivalence requires that he should be unable to tell that he (and the lift) are not in force-free space—but in that case, he certainly should *not* see any shift! Therefore, we must suppose that there is a compensating effect equal to a fraction v/c *upwards* in frequency, due to the fact that the gravitational potential is smaller (more negative) at the point of reception than at the point of emission. An observer who is stationary with respect to the earth will see this upwards shift, but not the downwards Doppler shift, and will therefore see the light 'blue-shifted' by a fraction $v/c = gd/c^2$. So we make the remarkable prediction that light emitted at a height d above the earth's surface will be observed at the surface with a frequency shifted upwards by an amount of approximately gd/c^2. Conversely, light emitted at the earth's surface will be observed at a height d above it with a red-shift of approximately the same amount—hence the name 'gravitational red-shift'. At a height of 1,000 feet, the fractional shift is only about 1 part in 3×10^{12}, but it has nevertheless been measured experimentally.

So far, we have implicitly assumed that the gravitational acceleration in the region of interest is uniform, in both magnitude and direction. As we have seen, under these conditions the principle of equivalence says that we can always go into a frame of reference (the freely falling one) in which things look just as they would in free space. In some sense the effects we have mentioned (curvature of the path of light and gravitational red-shift) could be regarded as an illusion due to our perverse choice of a frame of reference which is *not* freely accelerating under gravity (just as we tend to regard the apparent force exerted on a luggage-trolley—or ourselves—in a train which *is* accelerating along the earth's surface, but not under gravity, as illusory). However, if the acceleration due to gravity is not uniform in space or time, then it will not in general be possible to transform away all such effects by going into a locally freely falling coordinate system. Such a situation is bound to arise near a concentrated massive object, since the direction of gravitational acceleration is everywhere towards the centre of mass of the object, and cannot therefore be uniform in space. Under such conditions the general theory of relativity says that the presence of the massive object 'curves', or 'warps', the space–time continuum around it.

The idea of a curved space, or, worse still, a curved space–time, is a subtle and counter-intuitive one. A conventional way of explaining it goes something like this. If we compare the surface of a table with that of a football, we see at once that, by our ordinary everyday definitions, one is 'flat' and the other is 'curved'. But this is obvious to us only because we live in a three-dimensional space and can, for example, see directly that the football completely encloses a region of three-dimensional space. (More technically, we can see that the dependence on two of the coordinates of a point on the surface, say x and y, of the third coordinate, z, is quite different in the two cases.) But suppose that we were blind ants crawling on the surface of the table or the football, so that we had no concept of a third dimension. Could we tell whether our two-dimensional 'space' was flat or curved? In fact we could—for example, by the following experiment. We take a piece of string, fix one end to a particular point on the surface, attach a pencil to the other end, and, keeping the string stretched taut against the surface, move the pencil so as to trace out a circle. Then we untie the string and lay it repeatedly along the circle we have drawn so as to find out how many lengths of the string are contained in the circumference of the circle—that is, we measure the ratio of the circle's circumference to its radius. If we do this experiment on the flat table-top, the answer is of course 2π. If we do it on the surface of the football, however, the answer is always *less* than 2π, and depends on the length of the string; in fact, when the length of the string approaches half the circumference of the football, the circumference of the circle clearly approaches zero! If we were to do the same experiment on a saddle-shaped surface, the ratio would be *more* than 2π (such surfaces are said to have negative curvature). In this way we could infer not only whether our two-dimensional space was curved, but if it is, what the radius of curvature is.

Similar principles hold also in three-dimensional space, although it is much more difficult to form an intuitive picture in this case. It turns out that the familiar 'flat' Euclidean space we have always assumed, in which, for example, parallel lines never meet and the angles of a triangle always add up to 180°, is actually a special case of a much more general 'curved' space, which in principle could be locally of either positive or negative curvature, roughly corresponding to the football or the saddle in two dimensions.[3] (Remarkably, this possibility had been anticipated nearly a century before Einstein by the German mathematician Gauss, who took—at least according to anecdote—accurate measurements of the angles of a large triangle formed by three mountain peaks to check whether they added up to 180°—as in fact, to the accuracy of his experiment, they did.) An essential prediction of the general theory of relativity is that the curvature of space (more precisely, of space–time) depends on the distribution of matter (mass) in the neighbourhood; in particular, in the neighbourhood of a large concentrated mass (such as the earth or the sun), space has a slight negative curvature proportional to the mass. This effect is very small, the radius of curvature close to the sun being about 5×10^{11} metres—that is, about 700 times the actual radius of the sun itself. Despite this, the curvature is predicted to have observable effects: it is partly responsible for the well-known precession of the perihelion of Mercury—that is, the fact that the point in its orbit at which Mercury approaches most closely to the sun (perihelion) appears gradually to rotate around the sun relative to the fixed stars (the rest of this effect comes purely from *special* relativity)—and it also affects the actual magnitude of the light-bending effect discussed above. For masses much more concentrated than that of the sun, much more dramatic effects are predicted (see below).

The three 'classical' tests of general relativity—the gravitational red-shift, the bending of light in a strong gravitational field, and the precession of the perihelion of Mercury—all involve extremely tiny effects, and in each case there are alternative theories which could explain the data. But the conceptual beauty and simplicity of Einstein's theory are so overwhelming that ever since its formulation in 1916 it has been generally regarded as the prime candidate for a general theory of space, time, and gravitation. (In recent years, other tests besides the classical ones have been carried out, and have generally confirmed the theory.) If we accept it as generally valid, then we can extrapolate it far beyond the solar system, where it has been directly tested, to the cosmos as a whole; where this extrapolation has observable consequences—for example, in the dynamics of some binary stars—it seems at least consistent with the observations.

The models allowed by general relativity for the space–time structure of the universe are many and varied, and even now not all their properties have been fully worked out. However, there is a class of particularly simple models which seem compatible with what we presently know, whose properties have been well explored. These are the models in which the distribution of matter is homogeneous on the relevant scales. In this case it turns out that, given that the universe is currently expanding, there are three possibilities, which actually correspond to the three alternatives mentioned above in pseudo-Newtonian theory. First, the universe is infinite in spatial extent, has negative curvature, and is destined to expand for ever (in the sense that the apparent distance between any two galaxies will appear for ever to increase with time, and the rate will not gradually tend to zero). This is known as an 'open' universe. Second, the universe is infinite in spatial extent, but 'Euclidean' in geometry; that is, our everyday notions apply. In this case the expansion rate will gradually decrease towards zero as time goes on. This is called a 'flat' universe. Third, the universe is positively curved and finite in spatial extent, in roughly the same sense as a football, and will continue to expand only up to a maximum size, after which it will re-contract to a point. This is known as a 'closed' universe. The way in which the expansion behaves with respect to time (as observed by any observer moving with his local galaxy) turns out, rather surprisingly, to be identical with that predicted by pseudo-Newtonian theory; and as in the latter, the three cases correspond respectively to a present mass density below, equal to, and above the critical closure density of about 2×10^{-29} grams per cubic centimetre. As remarked earlier, the currently observable contents of the universe add up to a mass density of about 1 to 2 per cent of this figure, so that in the absence of dark matter the universe would be open and infinite.

Before leaving the subject of general relativity, let us look at two more predictions which it makes. One, which is actually not peculiar to Einstein's theory, but is shared by some other theories of gravitation, is the existence of gravitational waves. Just as in electromagnetism an oscillating electric charge produces electromagnetic waves, which travel away from the source with the speed of light, c, and then act on distant charges, so a suitably oscillating distribution of mass is predicted to produce gravitational waves (a deformation of the local space–time structure), which travel outwards at speed c and can in principle be detected by their effect on distant masses. The effect is very tiny, so that even a fairly violent supernova explosion in a relatively nearby star would produce waves which are detectable on earth only with extremely sensitive detectors; nevertheless, a number of experiments are currently in progress to try to detect such radiation.

A second, more spectacular phenomenon predicted by the general theory of relativity is the existence of black holes. The term 'black hole' conjures up alarming and sinister images (I once found a relatively sober book on the subject shelved in a bookshop under 'Science Fiction and Horror'!), but it in fact describes a spectacular warping of the space–time structure by a concentrated mass which is in itself fairly harmless. We have seen that if light is emitted by a stationary source in a low (negative) gravitational potential and is observed by an observer who is also stationary but in a higher potential, then the light he sees is red-shifted; this is equivalent to saying that if the source emits, by its own 'clocks', let us say, 10^{16} wave crests per second, the observer will see these 10^{16} crests arrive in somewhat more than a second; so to him, the source's clocks will appear to run slow. Now let us ask: If we could decrease the gravitational potential further and further—that is, make it more and more negative—could we red-shift the light seen by a distant observer to infinite wavelength? That is, could we arrive at a situation in which the observer thinks that the source's clocks have actually stopped? According to the general theory of relativity the answer is yes, and the gravitational potential needs to be equal to $-\frac{1}{2}c^2$ for this to happen. Since for an object of mass M and radius R, the gravitational potential is $-GM/R$, where G is the universal gravitational constant, we must have $R=2GM/c^2$; the radius R defined in this way is called the 'Schwarzschild radius'. For a body of the mass of the earth, it is about 1 centimetre, and for the sun about 3 kilometres, in each case much smaller than the actual physical radius of the body. Since the formula for the gravitational potential just quoted applies only when the mass is concentrated within a radius less than R, the Schwarzschild radius in these cases has no physical meaning. However, for a body with the density of a typical neutron star and a mass only slightly greater than that of the sun, the Schwarzschild radius is predicted to lie outside the physical radius. In this case, we find the startling result that if a spaceship (for example) moves towards the body in question, emitting signals with a given constant frequency as timed by the clocks on board, then, as received by a distant observer, these signals gradually decrease in frequency; and as the spaceship approaches the Schwarzschild radius, the frequency of the received signals tends to zero, so that the spaceship appears (to the observer) to take for ever to cross that point. On the other hand, an astronaut on the spaceship notices nothing particular happening when he crosses the Schwarzschild radius. He will go on emitting signals from inside it; it is just that they will never reach the distant observer. A picturesque way of saying this is that even light, once inside a black hole, cannot escape. (Notice, here, the connection with the light-bending phenomenon mentioned earlier: the Schwarzschild radius is just the point at which the radius of curvature of the light path becomes equal to the radius itself, so that the path of a ray tangential to the massive object is a circle. It is also amusing to note that if one thinks of light as a classical particle which travels at velocity c, and applies simple Newtonian mechanics, one reaches the same conclusion: namely, that once the light is inside the Schwarzschild radius, its kinetic energy, $\frac{1}{2}mc^2$, is insufficient to let it escape, and it stays trapped inside for ever. Indeed, the idea of a black hole was in effect formulated back in the eighteenth century by Michell and Laplace, by a similar argument.) It follows that we quite literally cannot 'see' anything inside a black hole (that is, inside its Schwarzschild radius), hence its name. If light cannot escape, then certainly nothing else can either.

Black holes are an inevitable consequence of extrapolating the formulae of the general theory of relativity, but do they exist in the real world? Although the evidence on this point is not conclusive, the general belief is that they probably do. A plausible candidate for a real-life black hole is provided by a star (Cygnus X-1) in the constellation Cygnus, which appears to be behaving as a member of a binary—that is, to be orbiting a companion star, which, however, is totally invisible by any ordinary means (including radio waves, X-rays, and so on). This companion could be a black hole. An even more exciting possibility is that the 'active centres' observed in some galaxies (see above) may be very massive black holes. Indeed, it is even conceivable that there is a massive black hole sitting at the centre of our own galaxy, no more than 30,000 light-years away from us. If that is so, it is conceivable that our remote descendants may be able to verify this directly.

Let us now turn from the present structure of the universe to the question of its history and genesis ('cosmogony'). Here, we are rather in the position of archaeologists trying to infer the structure of a vanished civilization from a few sherds and artefacts; whatever our view of what happened 'in the beginning', almost certainly most of the important objects which were produced at that time have long since vanished, or at least are beyond our ken, and we are left with at best a fairly random collection of debris. But, if we accept general relativity, or more precisely the simplest models based on it, and also our usual thermodynamics, we can make one general inference with some confidence: as we go back further and further into the history of the universe, it must have been not only progressively more dense, but also progressively hotter. As it expanded, it cooled down, like an expanding gas, and the present temperature of the cosmic blackbody radiation (about 3 degrees absolute) is the relic of very much higher temperatures in the past. In fact, most theories hold that in the first few instants of its existence the temperature of the universe must have been at least 10^{32} degrees. Now we know that if we take ordinary matter here on earth and heat it up, it progressively dissociates into more and more elementary constituents: solids melt, and the resulting liquids vaporize; molecules dissociate into their constituent atoms; the atoms in turn are progressively ionized until nuclei and electrons are separated; and eventually, if we could attain such temperatures, we believe that nuclei would dissociate into their constituent nucleons, and the nucleons in turn into their constituent quarks (and, who knows, there might be even further levels yet unexplored). So we arrive at a picture of the early universe as a sort of extremely hot, extremely dense 'soup' in which the primordial constituents of matter explode outwards from an initially infinite density. Such a 'scenario'—the vogue word for a qualitative picture which does not necessarily imply a quantitative realization—is usually referred to as the 'hot big bang'.

What kind of evidence can we produce for the hot big bang, apart from the fact that extrapolation of the equations of general relativity and thermodynamics seems to lead ineluctably back to it? One fairly convincing piece of evidence comes from the present relative abundances of the light elements, particularly hydrogen and helium and their isotopes, in the universe as a whole. Assuming, as always, the validity of our laboratory-tested laws of nuclear physics on the cosmic scale, we can calculate the amount of helium (and deuterium, the heavy isotope of hydrogen) which should have been produced at the stage of the big bang at which free nucleons began to condense into nuclei (the process known as 'nucleosynthesis'); the resulting abundance ratios seem to be in reasonable agreement with what is observed. If we go back beyond this stage, the situation is a bit more complicated, as we shall now see.

Probably the single most remarkable development in physics over the last fifteen years or so has been the increasingly fruitful interaction between particle physics and cosmology in devising scenarios of the very early universe. It has increasingly become realized that, if we have a reliable theory of how matter behaves at very high energies, and hence very high temperatures, we should actually be able to trace the evolution of the universe back in time far beyond the era of nucleosynthesis, perhaps right back to a time of the order of 10^{-43} seconds after its origin. At first sight one might think that speculations about such early epochs would be rash indeed. All our experience with ordinary matter suggests that the closer it is packed together, the more important the effects of interactions between the constituents (for a gas, the interactions between molecules can be completely neglected for many purposes, while for a solid, they are crucial); so one might think that the behaviour of matter at the incredibly high densities which occurred in the early universe would depend crucially on the details of the forces between the nucleons, quarks, or more elementary constituents, and is likely to be extremely complicated. What has changed the picture, thereby making the whole programme plausible, is the increasingly confident belief among particle physicists that the behaviour of matter at very high energies is actually *simpler* than that at lower energies; and in particular, that at high enough energies the constituents of matter in some sense cease to interact at all (the so-called asymptotic freedom described in Chapter 2). If this is right, then the universe in its initial stages might actually have been a much simpler place than one would have guessed.

As a consequence, there has evolved over the last ten years or so a picture of the early universe in which, initially, all possible symmetries were preserved and matter existed in the form of its most elemental constituents. As the universe expanded and cooled, there was a progressive breakdown of symmetry—first the (proposed) symmetry which unifies the strong and electroweak interactions, so that quarks and leptons were no longer interconvertible (this probably happened about 10^{-35} seconds after the big bang, at which point the temperature would have been about 10^{28} degrees), then the symmetry which unifies the weak and electromagnetic interactions (at about 3×10^{-7} seconds, at a temperature of about 10^{15} degrees); and finally, at about 10^{-5} seconds and 10^{12} degrees, the 'translational' symmetry was itself broken, that is, quarks and gluons ceased to form a uniform sea and started to form the nucleons we know (neutrons, protons, and various unstable particles, which rapidly decayed). Shortly thereafter, we enter the era of nucleosynthesis, on which we have rather more direct information (see above). The final picture is rather remarkable: it is in some sense as if we had to spend all our lives inside a regular crystalline solid, so that everywhere we went we saw the regular crystalline pattern and its consequences, and yet were able to infer that originally the atoms must have formed a gas or liquid without any such periodicity.

The details of this scenario are fluid, changing with time as experiments in particle physics confirm or rule out specific hypotheses regarding the way in which the symmetries are broken at each stage; but almost all versions have in common the feature that they predict that various kinds of debris (monopoles, cosmic 'strings', or domain walls) should be left over from the earlier stages, and should be observable today. (It is as if, by observing the defects in the crystal we are forced to inhabit, we could infer the way in which it had originally cooled from the melt.) At first sight, therefore, one would think that it should be fairly easy to check, by observing this debris, which of the detailed scenarios is correct. Alas, life is not that simple. First, it turns out that in many cases we have only a rather vague idea of what the debris in question would actually look like if it were around in our neighbourhood; indeed, in some cases, whether it would be observable at all. Second, a modification of the picture which has become popular in the last five years or so (the so-called inflationary universe scenario) predicts that at one or more stages in its early history the universe underwent a very rapid expansion, much more rapid than would be inferred from an extrapolation backwards of its present dynamics, and that as a result, most of the debris produced by the breaking of symmetries in its early history was removed to beyond the furthest horizon now visible to us—a truly cosmic case of sweeping unwanted entities under the rug! As a result, arguments for and against specific versions of the early universe scenario tend to be complicated and highly circumstantial; and at the time of writing it is not clear which is most likely to be correct.

Assuming that the scenario outlined above is in essence correct, why can we not follow it right back to the initial instant of the big bang? The reason lies at least in part in the absence, at present, of an agreed synthesis of quantum mechanics and gravity. As treated by the general theory of relativity, the gravitational field (or more accurately, the so-called space–time metric) is a field just like the electromagnetic field; however, while we have known how to quantize the electromagnetic field for nearly sixty years, the corresponding procedure for the gravitational field continues to elude us. Part of the reason is that whereas in electromagnetism the space and time coordinates are just labels of points at which the field is specified, and their measurement is not problematic, in general relativity the very measurement of position and time is itself determined by the field in question; in addition, there are special technical difficulties with gravitation. Thus, there is at present no agreed theory of 'quantum gravity'. Now, it turns out that this does not matter too much, provided that the fluctuations of the gravitational field are small compared with its average value (just as we can usually neglect quantum effects in the electromagnetic signals emitted by a radio station). This condition turns out to be well fulfilled provided that the scale of the phenomena in question is large compared to the so-called Planck length, a combination of the universal gravitational constant, Planck's constant, and the speed of light, which is about 10^{-33} centimetres. Now, as we go further and further back in time, the distance which light, and therefore any causal influence, could have propagated since the beginning of everything becomes smaller and smaller. When it falls below the Planck length, which happens at a time of about 10^{-43} seconds after the big bang, all bets are off: even those who believe we can trust the known laws of physics back to that point generally agree that to describe the universe in the first 10^{-43} seconds of its existence a fundamentally new approach will be needed. Indeed, quite apart from the quantum gravity problem, it is not even clear whether the concept of a 'beginning of the universe' (and hence, presumably, of a 'beginning of time') really has any meaning. All we can really say is that if we extrapolate backwards in time the equations which we believe govern the universe now, then we reach a point beyond which they make no sense. But this may say more about us than about the cosmos!

To sum up: just about everything we know, or think we know, about the universe and its history is based on extrapolation of the laws of physics as discovered in the laboratory to conditions different by many orders of magnitude in density, temperature, distance, and so forth. That we can get in this way a provisional picture which has even a reasonable chance of being self-consistent is little short of amazing. Even given this, the list of fundamental things we don't know about the universe is daunting. Among other things, we don't know what it is (mostly) made of, whether it is finite or infinite, whether it really had a beginning, and whether it will have an end. It is clear that we have a long way to go.

4
Physics on a human scale

If we were to take a census of all the known matter in the universe, the vast bulk of it would be either in the form of an incredibly dilute intergalactic gas (with a density of perhaps one atom per cubic metre—far less than we can achieve with the best vacuum techniques here on earth) or a constituent of stars of one kind or another, in which case it is usually very hot or very dense or possibly both. The fraction of matter which is at the right temperature and density to form solids, liquids, or even gases as we know them on earth is incredibly tiny—perhaps 1 part in 10^{11} of the whole. Yet, almost certainly, there are more physicists in the world working on matter in these phases—on so-called condensed-matter physics—than on all other areas of physics combined. Why so? There are a number of obvious reasons which, from the point of view of this book, are not particularly significant: the physics of condensed matter is what we are most used to doing, since most of our current knowledge of matter at the atomic level actually emerged from such studies; it is cheap and can be done by individual researchers or small groups, rather than the cast of hundreds which may be necessary for a particle physics experiment; and, needless to say, it is the main area of physics—in fact, by a liberal definition, the only area—which has the potential for major technological spin-offs. None of these reasons, however, bear directly on the *intellectual* challenge of condensed-matter physics.

Recently Sheldon Glashow, a leading particle physicist, responding to a suggestion that the funds required for the construction of the next ultra-high-energy accelerator might be put to better use if spread around other areas of science such as condensed-matter physics, conceded that there are 'important theories' in these other areas, but went on: 'How truly fundamental are they? Do they not result from a complex interplay among many atoms, about which Heisenberg and his friends taught us all we need to know long ago?'[1] The attitude implicit in these obviously rhetorical questions is a common one. It embodies at least two assumptions: first, that there are no new laws of nature to be discovered by studying condensed matter as such, since all behaviour of such matter follows, in principle, from the behaviour of its atomic or subatomic constituents; and second, that if this is so, then the study of complex matter cannot be as 'fundamental' (or, by implication, as interesting or deserving of support) as the study of the constituents themselves—indeed, that it is really rather a trivial occupation by comparison. I believe that both these assumptions are questionable. To question the first implies a somewhat unorthodox point of view, which I shall touch on in the last chapter; in the present context let me simply concede it for the sake of argument, and examine the second.

In what sense can we say that the study of individual particles (be they atoms, nucleons, quarks, or whatever) is more or less fundamental than the study of their collective behaviour in a solid or a liquid? An analogy from the social sciences may be useful. Very few people indeed would seriously maintain that it would be realistic to expect to understand the social psychology of (say) nations merely in terms of observations of the behaviour of individuals, or even pairs of individuals, studied in isolation from the rest of the society (let us say, on a desert island 'laboratory'). One obvious reason is that the very interaction between a pair of individuals, if it is at any level of subtlety, is likely to be itself profoundly influenced by the social environment in which the interaction takes place. Accordingly, sociologists and social psychologists have developed theories of collective human behaviour which are in some sense autonomous, and which do not refer (or not necessarily) to the kind of information one could gain from the study of an isolated individual, or even from the interactions of pairs or small groups.

If such an approach is necessary when studying the collective behaviour of the 10^8 or so human beings who make up a nation, for example, why should it be any less necessary in the study of the 10^{16} or so atoms that make up a speck of dust, let alone a small biological organism? There is, to be sure, a widespread prejudice that physics is indeed different in this respect, that it has progressed precisely because we have analysed the behaviour of complex matter into that of its constituent particles. (Indeed, some social scientists seem to hanker after this alleged 'reductionist' aspect of physics as a model for their own disciplines.) This is at best a partial truth. Of course, there are some cases in which the behaviour of a macroscopic system (a liquid or a solid, say) is indeed reducible to that of its constituent elements. For example, if we neglect the effects of special relativity, then the mass of a piece of iron is just the mass of the atoms composing it; and the dielectric susceptibility of a piece of solid argon is to a good approximation equal to the sum of the susceptibilities of the constituent argon atoms. These examples are rather trivial, however; they are analogous to the statement that the total food consumption of a nation is equal to the summed consumptions of the individuals composing it—a remark which does not tell us a lot about the nature of social psychology. Rather less trivially, there are also a number of cases, particularly in the traditional areas of the physics of gases and crystalline solids, in which a model which treats the behaviour of the whole as essentially just the sum of that of its parts (atoms or electrons) has been quite successful; and a few more in which, even if a 'one-particle' picture fails, a description in terms of pairs of particles interacting in a way which is not particularly sensitive to the environment gives good results. But these cases, despite the fact that they totally dominate the presentation of the subject in most elementary textbooks, are actually the exception rather than the rule. In virtually all the frontier areas of modern condensed-matter physics, the relationship between our understanding of the behaviour of matter at the microscopic level of single atoms and electrons, and at the macroscopic level of (say) liquids and solids, is actually a good deal more complicated than this.

Even granted, though, that the interactions between the elementary constituents of matter may be all-important in determining the behaviour of a macroscopic body, is it not true that we can work out the consequences of these interactions? It seems to be a common belief that all condensed-matter theory essentially consists in writing down specifications of the microscopic entities which compose a body, and of their interactions, applying the fundamental laws which we believe to govern the behaviour of the microscopic entities (Newton's laws, if we are doing classical mechanics; Schrödinger's equation if we are doing quantum mechanics), and then, by a process of purely mathematical derivation or 'approximation', deducing the behaviour of the macroscopic body in question. Were this correct, condensed-matter physics, or at least the theoretical side of it, would indeed be a somewhat boring and, in the literal sense of the word, 'derivative' subject. In fact, however, this picture is not only false but, were it true, would describe a totally arid occupation irrelevant to any real problems. Let me take the second point first. In the first place, whereas in a particle-physics context the information that we have a beam of, let us say, protons which have been accelerated to a given energy and propagated in a given direction with a given (spin) polarization is sufficient to characterize the beam and its constituents completely, in condensed-matter physics we never in practice know all the details which we would need in order to write down a complete microscopic description. For example, all crystals contain chemical impurities to some degree or other, and while we often know the average concentration of impurities, we rarely, if ever, know how they have chosen to distribute themselves in the crystal. In the second place, whereas our beam of protons travels in a high vacuum and can be treated to a very good approximation as isolated from external influences, a macroscopic body is, at least under ordinary laboratory conditions, continually and (usually) strongly interacting with its environment—if in no other way, through the walls of the container holding it or the laboratory bench on which it sits, not to mention the blackbody radiation field. Unless we are willing to get into an infinite regress, it is necessary to specify the effects of this interaction in a way which does not require the writing down of a detailed microscopic description of the environment—and thus to spoil what was supposed to be the main point of the exercise. In the third place, let us suppose for the sake of argument that with the advent of currently undreamed-of computing power it becomes possible in the future—as it is certainly not today—to solve Schrödinger's equation for the 10^{23}-odd atoms comprising a typical liquid or solid for one choice of microscopic description and initial conditions, or even perhaps a variety of such conditions. What would we have learned? Almost certainly, nothing. We would be faced with a pile of millions of tons of computer print-out (or, more optimistically, computer graphics) which would certainly 'in principle' contain the answer to any question we might want to ask, but in a form which would be totally useless to us without some consciously chosen principles of organization. Such an outcome is certainly not what we mean by 'understanding' the behaviour of matter in bulk.

If the activity just described is not what condensed-matter physics is all about, then what *is* it about? I would claim that the most important advances in this area come about by the emergence of qualitatively new concepts at the intermediate or macroscopic levels—concepts which, one hopes, will be compatible with one's information about the microscopic constituents, but which are in no sense logically dependent on it. These new concepts may range from the highly abstract—voltage, entropy, correlation length, and so on, to pictorial models which are substantial aids to our quasi-visual intuition (as, for example, when the interactions of atoms in a solid are modelled by microscopic springs); what they have in common is that they provide a new way of classifying a seemingly intractable mass of information, of selecting the *important* variables from the innumerable possible variables which one can identify in a macroscopic system; in the language of psychology, they present a new *Gestalt*. Many of the older concepts of this type are so deeply embedded in the language of modern physics that it is difficult to imagine the subject without them. What is perhaps less generally realized is that this process of *Gestalt* change (or better, *Gestalt* expansion) did not stop at the end of the nineteenth century but is going on actively today, admittedly usually at a rather more material-specific level.

All this is not to deny that an important role is played in condensed-matter physics by attempts to relate the macroscopic behaviour of bulk matter to our knowledge concerning its constituent atoms and electrons. Indeed, the theoretical literature on the subject is full of papers which at first sight seem to be claiming to 'derive' the former from the latter—that is, to do exactly what I have just said condensed-matter physicists do *not* do. The question of what, in fact, such papers are really doing is a fascinating one, which in my opinion is far too little studied by philosophers of science, but it would take me too far afield to discuss it in depth here, so let me just make two remarks. The first is of the nature of a well-kept trade secret: the word 'derivation' as used in these papers often bears a sense quite unrecognizable to a professional mathematician. In fact, it is often used to describe a hybrid process, in which some steps are indeed mathematically rigorous, while others—so-called 'physical approximations'—are actually not approximations in the usual sense at all, but rather are more or less intelligent guesses, guided perhaps by experience with related systems. In other words, according to the distinction made in Chapter 1, mathematical and physical arguments are intimately intertwined, often without explicit comment. The second remark is that, in those relatively rare cases where one does indeed derive the behaviour of a macroscopic body from a 'microscopic' description in a way which a professional mathematician might find acceptable, the microscopic description in question is usually itself only a 'model' or representation which, while it encapsulates (or so we hope) the *relevant* features of the system in question, is often very far indeed from what we believe to be the true description at the microscopic level. (An example of such a model is given below, where I discuss second-order phase transitions.) It is precisely this compelling need to isolate, from a vast and initially undifferentiated mass of information, the features which are relevant to the questions one wishes to ask, which distinguishes condensed-matter physics qualitatively from areas such as atomic or particle physics. (And which, incidentally, could be argued to make it a much more appropriate paradigm than them for the social sciences, should such a paradigm be felt to be needed.)

In this situation I believe that it is sensible to reorient our view of the kinds of questions that we are really asking in condensed-matter physics. Rather than chasing after the almost certainly chimerical goal of *deducing* the behaviour of macroscopic bodies rigorously from postulates regarding the microscopic level, it may be better to view the main point of the discipline as, first, the building of autonomous concepts or models at various levels, ranging all the way from the level of atomic and subatomic physics to that of thermodynamics; and, second, the demonstration that the relation between these models at various levels is one not of deducibility but of *consistency*—that is, that there are indeed 'physical approximations' we can make which make the models at various levels mutually compatible. From this point of view, indeed, the most important theorems in physics are theorems about the *incompatibility* of models at different levels; one such is the well-known theorem of Bohr and van Leeuwen, which states that no model of atoms which employs only classical mechanics and classical statistical mechanics can produce the observed atomic diamagnetism; and we shall meet another famous example (Bell's theorem) in the next chapter.

Let me conclude this discussion with an analogy. Suppose one wanted to represent in compact form the transportation network of a country like England. There are at least two ways one could go about it. On the one hand, one could take an extensive series of aerial photographs and reduce them in scale. In this way one would in some sense have started with an 'exact' picture and made 'approximations' to it; the degree to which objects would be accurately represented in it would be a function simply of their physical size, and would in no way reflect their intrinsic importance in the communications network. Such a picture would be of little use as a practical guide. A much more useful alternative would be simply to draw a road or rail map of the type with which we are all familiar—that is, something which in no sense pretends to be a picture, but rather is a schematic representation of that information which is important in the context. Naturally, the map must be *consistent* with the topography shown in the detailed aerial photographs, but it is in no sense an *approximation* to them; indeed, it may embody quite different kinds of information (for example, British road maps conventionally represent roads by different colours in accordance with the Ministry of Transport classification, something which no amount of inspection of aerial photographs could reveal!). The view which I have been arguing against (which seems to be widespread among philosophers of science and other onlookers, and more surprisingly even among some practising scientists) in effect holds that the theories and concepts of condensed-matter physics are analogous to the reduced aerial photographs; my claim is that they are much better viewed as analogous to the map.

The subject-matter of condensed-matter physics is by its nature extraordinarily various, ranging from the behaviour of liquid helium at temperatures below one-thousandth of a degree to the physics of a possible melt-down in a nuclear reactor core, from the electronics of ultra-small silicon chips to the psychophysics of human vision. From this vast field I will simply pick out a few of what I would regard as the 'frontier' problems and say a little about them. Before doing so, however, I want to take a minute to ask how we actually get our information.

As we saw in earlier chapters, we obtain our experimental information about the distant reaches of the universe exclusively by waiting for photons or other particles emitted there to arrive on earth or nearby; in particle physics, on the other hand, the vast bulk of our information comes from a single type of experiment, in which we fire one type of particle at another type and watch what comes out and in which direction. In investigating condensed systems we have available both these types of experiment and a great deal more besides. In fact, we can actually obtain quite a lot of useful information from experiments which are analogous to the accelerator experiments of particle physics; in this case the target is a macroscopic body, which is almost always fixed in the laboratory, at which we fire elementary particles such as photons (visible light or X-rays), neutrons, electrons, protons, or pions, or even composite particles such as atoms of hydrogen, helium, or heavier elements. The basic experimental data are the same as in particle physics—namely, the number of particles scattered per second in a given direction with given energy loss; from this, we can infer a great deal both about the static structure of the material studied (as, for example, in a typical X-ray diffraction experiment) and about its dynamics. However, there are all sorts of other experiments we can do, many of them involving only *macroscopic* variables: we can apply a given amount of heat and from the temperature rise infer the specific heat; we can apply a voltage and measure the total current flowing through the body (thereby measuring its resistance); and so on. Again, we can do experiments involving substantial time delays; for example, we can excite the system using an intense short pulse of light, and then, by scattering light of a different wavelength from it later, watch the way in which it relaxes back to equilibrium. (This technique is particularly informative for biophysical systems, which can have very long relaxation times.) We can study the behaviour of (say) a carbon–iron alloy as a function of its history—for example, whether it cooled slowly from a high temperature or was 'quenched' by being plunged into a bath of cold water—and so on and so forth, the list of possible types of experiment being practically endless. In fact, part of the expertise of the experimental condensed-matter physicist lies in deciding which of the myriad types of experiment available to him is most likely to provide the information he is seeking, or reveal qualitatively new features of the material under investigation.

Experiments designed to measure the properties of a macroscopic system differ in at least two important respects from the kind of experiments carried out in particle physics. In the first place, as has already been pointed out, it is absolutely essential to specify the effects one thinks the environment is having on the system. Commonly, this is done in an 'averaged' way, by means of the thermodynamic concepts of temperature, pressure, and so on. It may also be necessary to specify other quantities, such as the magnetic field, which can have drastic effects on a system's behaviour. (By contrast, in considering the actual collision processes between elementary particles—as distinct from their motion between collisions—magnetic fields of the order of magnitude found in the laboratory are almost invariably negligible.) A second important difference concerns the reproducibility of individual 'events'. If one is doing a scattering experiment in particle physics, or for that matter in condensed-matter physics, the scattering of any one individual photon, proton, or whatever is a random event governed by statistical laws; only the distribution, obtained by summing a great many individual events, is reproducible, and in principle—even if not always in practice—it is possible to recover, in the data, the original randomness. In most condensed-matter experiments other than scattering ones, by contrast, one is measuring a quantity (temperature, pressure, resistance, magnetization, and so on) which by its nature is automatically a sum or average over many microscopic events, so that one expects to get a well-defined result which is reproducible apart from fluctuations which can be made as small as we like by making the system larger. Although this description fits the vast majority of such macroscopic experiments, there are important exceptions. For example, the temperature at which a supercooled liquid freezes is not always reproducible between one 'run' and the next, however carefully controlled the impurity content may be. This is probably a special case of a larger class of phenomena in which the macroscopic behaviour is pathologically sensitive to tiny changes in the initial conditions (see below). Again, in recent years experiments have been done in which the behaviour of a macroscopic body appears to be controlled rather directly by intrinsically quantum-mechanical processes, with all their attendant built-in indeterminacy.

There are three important types of condensed-matter systems for which we can claim a good quantitative, as well as qualitative, understanding, and these constitute, to an overwhelming degree, the subject-matter of textbooks in this area—to such an extent that one might easily get the impression that they exhaust the subject. These three are dilute gases, simple liquids, and perfect crystalline solids. In a dilute gas, by definition, the molecules are spaced far apart; for example, in air under ordinary room conditions, the average volume available per molecule is about 10,000 cubic ångströms, whereas the volume of the molecule itself, in so far as it can be defined, is only 2 or 3 cubic ångströms. As a result, the molecules move for long periods without coming near enough to their neighbours to be substantially affected by them (a molecule in air typically travels 10^{-4} centimetres—a huge distance by atomic standards—before making a collision). Consequently, an excellent first approximation to the behaviour of the gas is obtained by neglecting the interactions entirely, and the resulting theory is in quite good agreement with experiment as regards thermodynamic properties such as specific heat, compressibility, and so on. Some other properties, such as thermal conductivity, cannot be calculated in this way—indeed, they turn out to be infinite on this model—and to get a sensible description of these, it is necessary to take the interactions and collisions between molecules into account. However, when two molecules in a dilute gas collide, the chance of a third molecule being near enough to affect the collision is very small; so a theory which regards the two colliding molecules as interacting just as if they were completely isolated usually gives very good results. A theory of this type is said to ignore three- (or more) body correlations, and we can say that such a theory gives a very good quantitative description of the behaviour of most gases when they are sufficiently dilute.

In a typical liquid, on the other hand, the molecules are packed closely together, almost as closely, in fact, as in a solid.[2] However, unlike in a solid, the molecules are free to move around one another, and if their shapes and the chemical forces between them are reasonably simple, then this motion can actually occur without much hindrance, right up to the point at which freezing takes place. The molecules may be pictured as rather like passengers in a moderately crowded tube train at rush-hour: for any one molecule to move may require a little accommodation by its immediate neighbours, but molecules more than a few ångströms away are hardly aware of the disturbance. As a result, although a theory which treats only 'two-body' correlations—that is, which treats the interaction between any two molecules as if it were taking place in vacuum—may not be sufficient, it is not necessary to worry about correlations involving large numbers of molecules (more than four or five, say). A further simplifying feature is that, with the exception of helium, and to some extent hydrogen, the liquid phases of all elements and compounds exist only at temperatures high enough that the effects of quantum mechanics on the molecular motion can be neglected to a good approximation. Thus, it is not usually thought that subtle quantum correlations of the type we shall mention in the next chapter play any role in such systems, and a rather simple classical picture—in fact, in some cases, a picture of the molecules as behaving simply like a collection of billiard-balls—gives qualitatively correct results. Actually, we can often do a lot better than this, with the help of modern computing facilities; while the quantitative agreement between theory and experiment for the properties of simple liquids is not of the same order as it is for dilute gases, it is nevertheless quite impressive.

In both dilute gases and simple liquids, one has, therefore, two simplifying features: the effects of quantum mechanics are negligible (except in gases at extremely low temperatures), and one can treat all interactions 'locally'—that is, neglect the effects of distant parts of the system. Neither of these simplifications applies in a crystalline solid, and the achievement, despite this, of a theory of the behaviour of such solids was the first and one of the most impressive triumphs of quantum mechanics as applied to condensed systems. The crucial feature of a crystalline solid is that the atoms, or rather the ions (the original atoms minus their outermost electrons), are arranged in a regular periodic array, like a three-dimensional wallpaper pattern. Of course, this is never exactly the case for any real solid—there are always imperfections of one kind or another—but corrections to this model are usually treated as an afterthought. If we consider only small displacements of the ions from their original, equilibrium positions, the forces between them are much as if they were joined by simple springs; so at least within the framework of classical mechanics we have a simple and intuitive mechanical model for the system. One interesting property it possesses is that if a single atom is moved, other atoms will feel the effect, however far away they are; in fact, the 'natural' modes of motion of the system propagate through the array of atoms like waves. (A one-dimensional analogue of this behaviour may be observed in everyday life when a locomotive shunts a long line of freight cars.) Indeed, if the wavelength of the disturbance is long enough, and we observe only on a scale corresponding to many interatomic distances, then what we see is a periodic compression and rarefaction of the solid—a sound wave. (Sound waves can propagate, of course, in gases and liquids as well as in solids, but in these cases the mechanism of propagation normally involves a very large number of random collisions, rather than the microscopically regular behaviour observed in crystalline solids.) Such a sound wave has, at any rate at long wavelengths, a frequency, ν, which is inversely proportional to its wavelength, and in this and other respects resembles an electromagnetic wave. It is not surprising, then, that when we apply the ideas of quantum mechanics to this situation, we find that just as the energy of an electromagnetic wave arrives in chunks (photons), so does the energy of a sound wave; the chunks (quanta) in this case are called 'phonons', from the Greek word for 'sound'. By applying the principles of statistical mechanics to the phonons, we can find out how many there are, on average, in the solid at a given temperature, and hence calculate thermodynamic properties such as the specific heat; generally speaking, very good quantitative agreement with experiment is obtained. What is particularly interesting about the idea of a phonon is that it is a genuinely collective phenomenon and depends crucially on the relative behaviour of atoms at large distances from one another; there is no way we can obtain the correct low-temperature behaviour of the specific heat of an insulating solid by considering only small groups of atoms separately. Despite this, the phonons themselves behave in many ways very like simple particles—and indeed are members of a class of entities usually called 'quasiparticles'.

An even more interesting application of quantum mechanics emerges when we consider the behaviour of the electrons in a crystalline solid (remember, the ions were conceived as having surrendered their outermost electrons). Since the ions are unable to move far from their lattice positions, it is presumably the electrons which carry any electric current flowing in the solid, and are thereby responsible for the electrical conductivity. Now, the electrical conductivity of a solid may be, in the appropriate units, anything from about 10^9 (very pure copper at low temperatures)[3] to 10^{-15} (for example, polyvinyl chloride)—a range of 24 orders of magnitude. How is this enormous range of behaviour to be understood? The oldest theory of conductivity, which took into account neither quantum mechanics nor the effects of periodicity, predicted in effect that all solids should be good conductors, and it turned out that the introduction of quantum mechanics alone did not change this picture. It is the *combination* of the effects of periodicity and quantum mechanics which is crucial. In brief, what happens is this: since, in quantum mechanics, the electron is regarded as a wave, it suffers diffraction from the three-dimensional 'diffraction grating' formed by the periodic lattice. In such a situation it turns out that the allowed frequencies of the wave, and hence the allowed energies of the electron states, form 'bands' separated by regions of frequency (energy) in which no wave can propagate. (Some readers may be familiar with the analogous phenomenon of 'pass bands' and 'stop bands' which occur in periodic arrays of electrical circuit elements—systems which are usually treated purely classically.) Now, the Pauli exclusion principle allows only one electron to occupy each state, and so the available states will be filled from the bottom upwards. It turns out that (in the simplest cases, at least) the number of states per band is just twice the number of ions in the lattice, and therefore the last state filled lies either just at the top of an allowed band or half-way up (depending on the number of electrons originally surrendered by each ion). Now, since in the original, equilibrium state of the system no current flows, it is at least plausible to think that if we wish a current to flow in response to an applied voltage, we must arrange for some of the electrons to move into new states. In the case where the last state filled lies half-way up a band, this is quite easy; it takes very little energy to move some of the electrons up into unfilled states, and an external voltage can supply this. If, on the other hand, the last filled state lies at the top of a band, then to populate the next state available requires a large energy, which, it turns out, under normal circumstances, cannot be supplied by an external voltage source. Thus, in the first case the solid readily carries current—that is, behaves as a good conductor (metal)—while in the second case it is a good insulator; a quantitative consideration shows that the conductivities in the two cases can indeed differ by many orders of magnitude.

The 'band theory of solids' outlined above is a one-particle theory—that is, it is based on the idea that it is adequate to consider each electron as moving freely through the crystalline lattice, without substantial qualitative effects from all the other electrons (except in so far as one must apply the Pauli exclusion principle). With some relatively minor modifications, it has been enormously successful not only in explaining the metal–insulator distinction, but in accounting quantitatively for most of the properties of most metals. This success is actually rather puzzling, for it is clear that the band theory cannot be the whole story. Were it so, it would be impossible to explain, for example, why many liquids, in which ions certainly do *not* form a crystalline lattice, are nevertheless insulators. Moreover, in the last four years or so the confidence of condensed-matter physicists in the band picture of metals has received a severe shock. It had long been appreciated that there were some phenomena involving phase transitions, such as magnetic ordering and superconductivity (on which, more below), which the band picture is clearly incompetent to explain and for which even a qualitative understanding requires an account of the interactions between electrons. However, there was a certain general confidence that in the absence of such phase transitions the band picture would not be qualitatively misleading, at least. In the last few years this expectation has been rudely shattered by the discovery of a large class of metallic compounds—the so-called heavy-fermion systems—which, even when not magnetically ordered or superconducting, show properties quite different—sometimes even qualitatively so—from those of familiar metals. While there is general agreement that these anomalous properties must reflect a serious breakdown of the simple band picture, there is at present agreement on very little else; and the study of these compounds is among the most active and exciting areas of contemporary condensed-matter physics.

If you read a typical textbook on solid-state physics—condensed-matter physics as applied to solids—you are liable to come away with the impression that anything you meet in real life, if it is not a dilute gas or a simple liquid, is likely to be a crystalline solid. Yet, if you glance around the room in which you are reading this book, you may well have considerable difficulty in identifying anything which is obviously crystalline. As I look around as I write this, I see wooden furniture (which is certainly not crystalline), books (ditto), a letter-tray made of some synthetic plastic (a nasty tangle of polymer molecules), a cup containing coffee (even the water is only dubiously a 'simple' liquid, and the organic molecules composing the coffee and milk are certainly much too complicated to qualify), and of course my own body, whose microscopic constitution is about as far from any of the systems we have discussed as can be imagined. In fact, just about the only constituent of my environment which even approximately qualifies for treatment under any of the three heads discussed above is the air. The simple fact is that in real life crystalline solids are very much in the minority, and the only reason that most textbooks of solid-state physics concentrate almost exclusively on them is that until quite recently they were the *only* kind of solids of whose behaviour we could even begin to claim a quantitative understanding. But this is the case no longer. Indeed, with the exception of the heavy-fermion systems mentioned above and a few other topics, most of the more exciting advances in condensed-matter physics over the last two decades have concerned either structures which are not crystalline, or problems where the crystallinity, if any, is of no great relevance. I will now sketch a few of these.

One problem where our understanding has advanced considerably in the last few years is the properties of very disordered materials, particularly their electrical conductivity. As we saw above, with some provisos, the behaviour of electrons in perfect crystalline solids is believed to be fairly well understood. If we now insert a few randomly distributed chemical or metallurgical impurities, the traditional approach would be to treat them as scattering the electron waves in such a way that different scattering events are more or less independent (just as, in a dilute gas, the successive collisions of a particular atom with two other atoms are regarded as independent events). But as we put in more and more impurities, we eventually reach a stage at which there really is no crystalline lattice left, and the whole picture becomes dubious. (Such a situation can be achieved in practice by quenching—cooling very rapidly from the liquid phase—certain types of metallic alloy; the atoms simply do not have time during the freezing process to find their optimum (crystalline) configuration.) So one really has to start again from scratch, and treat the electrons as moving in a very complicated and irregular field of force due to the randomly located atoms. The outcome is surprising: one finds that a sufficient degree of disorder causes the electrons to become 'trapped' in restricted regions of space, and this shows up in experiments as a transition from a metallic to an insulating state. This is a purely quantum-mechanical effect, with no obvious classical analogue; moreover, it is a 'global' rather than a 'local' effect, in the sense that it cannot be deduced by considering the scattering of the electrons by one or a few atoms at a time. It is still a 'one-electron' effect, however, in that it is possible to obtain at least a qualitatively correct result by treating the electrons as moving independently in the random force field. It is amusing that, whereas in the traditional textbook approach to solid-state physics the occurrence of insulating behaviour required the postulate of perfect crystalline order, it now turns out that sufficient *disorder* will have the same qualitative effect! In addition, it turns out that electron–electron interactions can sometimes produce an insulating transition even in the absence of disorder. However, we still do not have a complete understanding of the conditions under which a given non-crystalline substance will be metallic or insulating, especially in the case of liquids.

One general class of phenomena in condensed-matter physics in which the interactions are certainly all-important is that of phase transitions. The term 'phase transition' has a well-defined technical meaning, but for present purposes we can take it simply to refer to the process of conversion between two qualitatively different states, or 'phases', of a macroscopic body—for example, the liquid and the solid state, or in the case of a magnetic material the 'paramagnetic' phase, in which the substance possesses no magnetization in the absence of an external magnetic field, and the 'ferromagnetic' phase, in which such a spontaneous magnetization exists. There is an important distinction to be made between so-called second-order, or 'continuous', phase transitions, in which one phase can go over smoothly and continuously into the other, and first-order, or 'discontinuous', transitions in which the qualitative difference between the two phases cannot be made arbitrarily small. Crudely speaking, second-order transitions tend to take place between a 'disordered' phase and an ordered one, first-order transitions between two ordered phases with different types of ordering. The transition between the paramagnetic and ferromagnetic phases of a magnetic substance is usually second-order, because the amount of spontaneous magnetization can be made as small as we like; in fact, if we take (say) a piece of a metal such as iron which is ferromagnetic at room temperature and heat it slowly in the absence of a magnetic field towards the temperature T_C at which the transition to the paramagnetic phase takes place (the so-called Curie temperature), then, as we approach this temperature, the amount of spontaneous magnetization tends smoothly and continuously to zero. The amount of magnetization—though not its direction (see below)—is uniquely defined at all temperatures, and in particular is zero above the Curie temperature; hence it is impossible to have the ferromagnetic phase above T_C or the paramagnetic one below it. By contrast, the liquid–solid phase transition is first-order, at least when the solid in question is crystalline, since there is a fundamental qualitative feature which distinguishes the solid from the liquid, which cannot be made arbitrarily small: namely, in the solid phase the atoms are tied down near their lattice positions, while in the liquid phase they are free to move over arbitrarily large distances.

Because of this feature, the behaviour of a system as it goes from solid to liquid, or vice versa, may be much more complicated than in the case of the paramagnetic–ferromagnetic transition. The general principles of thermodynamics tell us that a system maintained at a given temperature and pressure will always try to lower its so-called free energy (which depends on its energy, entropy, and so on). If we think of the possible states of the system as analogous to horizontal points on the earth's surface, and the free energy as analogous to the height of the ground, that means that the system will always tend to 'roll downhill'. However, this does not mean that it will necessarily reach the lowest—that is, most thermodynamically stable—state; it may become trapped in a 'metastable' state (see below), just as water can be trapped in a hollow at the top of a hill, even though much lower states are available to it elsewhere. By adjusting the temperature, for example, we change the relative height (free energy) of the states; for example, at high temperature (greater than T_0, say), the state of lowest free energy is the liquid state; while at low temperatures ($< T_0$), the solid state will have a lower free energy. However, if the system starts out in the liquid state at high temperatures and is then cooled through T_0, it may be left, like water at the top of a hill, in a state which is metastable, that is, which is a local, but not an absolute, minimum; to make the transition from the (now only metastable) liquid state to the (now stable) solid state may require it to pass through states which have a higher free energy than either. Under these conditions it is possible for the system to remain in the liquid state, for a time at least, even when the temperature is below the temperature T_0 at which this phase is no longer the most stable—a phenomenon known as 'supercooling'. (Under certain circumstances the reverse can also happen—that is, the solid phase can be 'superheated'.) The way in which the transition finally takes place (if it does!) is a matter of great theoretical and experimental interest (and also of great practical significance in many contexts).

Let us now return to the problem of second-order phase transitions. This is perhaps the paradigm case, in the physics of condensed matter, in which one must take into account 'many-body' effects in a way that can never be reduced to the interactions of a few particles in isolation.[4] A crude argument to show why this is so might go as follows: as we approach the second-order phase transition temperature from either above or below, the difference in free energy per unit volume between the ordered phase and the disordered one becomes very small. Now, according to the general principles of statistical mechanics, a given subvolume of the system can fluctuate into a state which is not strictly its equilibrium state, provided that the extra free energy needed is of the order of the characteristic thermal energy or less. As we approach the transition temperature, therefore, larger and larger subvolumes can fluctuate between the ordered and disordered states, and because they are so delicately poised between the two possible types of behaviour characterizing those states, even very small influences from outside the subvolume can have a decisive effect. Thus, to determine the behaviour at a given point in the system as the transition is approached, we have to take into account influences from further and further away, or, equivalently, to consider the interactions of greater and greater numbers of particles: any theory which attempts to describe the behaviour of a given atom in terms of the effects of only its immediate neighbours is guaranteed a priori to fail under these conditions. This is true despite the fact that the basic mechanism by which the effects are propagated is still (in most cases) by interactions between pairs of particles at a time.

The last two decades have seen a remarkable advance in our quantitative understanding of the behaviour of systems near a second-order phase transition—an advance which has come, in a sense, precisely by welcoming and exploiting the fact that as the transition is approached, larger and larger numbers of particles effectively influence one another. One crucial realization has been that, since, close to the transition, it is the interaction of very large, almost macroscopic, subvolumes that is essential, the 'critical' behaviour—by which is meant such things as the general shape (rather than the scale) of the curve of magnetization versus temperature as the transition is approached—can be influenced only by those characteristics of the system which can sensibly be defined at the coarse-grained, macroscopic level. There are only two such characteristic properties: the spatial dimensionality of the system in question (which is normally 3, of course, but can, for example, be 2 in the case of phase transitions taking place only in a surface layer of atoms), and the kinds of 'symmetry' associated with the parameter which characterizes the ordered phase. For example, in a magnetic phase transition the magnetization might be constrained to lie in the positive or negative direction along a particular axis (the 'Ising model'), or alternatively it might be able to point in any direction in space (the 'Heisenberg model'); and these two possibilities could give rise to quite different quantitative behaviour near the second-order transition. On the other hand, if we consider a transition occurring in a set of asymmetric rod-like molecules in a liquid, in which the molecules must all order themselves so as to be parallel to one another, but where their common orientation can be in any direction, the symmetry is exactly the same as that of the

Heisenberg model of the magnet, and the critical behaviour of the two systems is predicted to be the same: they are said to belong to the same 'universality class'. All features of the model other than dimensionality and symmetry—the exact nature and magnitude of the forces between neighbouring atoms, and so on—can affect the value of the transition temperature and the overall magnitude of thermodynamic and other quantities near it, but not their critical behaviour.

It is precisely this feature of universality which has made it reasonable and fruitful to study various simplified models of real systems, even though one is conscious that in many respects such models are very bad descriptions. For example, one has some confidence that the crude model of a magnetic material such as iron which is obtained simply by imagining microscopic magnets placed on the lattice sites, provided only that the symmetry is chosen correctly (that is, that the magnets are allowed to point along only one axis or in any direction in space or something else, as appropriate), will give the critical behaviour correctly, even though it may fail for just about any other purpose. (For example, it is quite hopeless for predicting the scattering of X-rays from the material in question.) The theory of second-order phase transitions is thus one area of condensed-matter physics—one of rather few, actually—where it is indeed possible to start from a well-defined model and do what is essentially a calculation in applied mathematics with some confidence that the results will have something to do with real life.

Before we leave the subject of second-order phase transitions, we should return briefly to a concept that was introduced in Chapter 2—namely, spontaneously broken symmetry. We continue to consider, for the sake of concreteness, a ferromagnetic phase transition, and to suppose that the interactions in the system in question are described by the Heisenberg model. This means that the energy is lowest when all the spins lie parallel to one another, but that no particular common orientation is preferred to any other; thus, as noted above, while the magnitude of the total magnetization is fixed, it can point in any direction in space. In practice, small residual effects, such as anisotropy of the crystal lattice or, failing that, the earth's magnetic field, break the symmetry and induce the magnetization to point in a particular direction; but it is still very useful to consider the idealized case in which there are no such residual effects and the direction is therefore completely arbitrary—in which case we say that the symmetry is spontaneously broken. The concept of spontaneously broken symmetry can be applied to a wide variety of ordered phases, even in cases where the transition is not second-order; for example, the regular pattern formed by the atoms in a crystalline solid is a case in point, the symmetry which is broken in this case being the 'translational' symmetry, that is, the symmetry under a simultaneous uniform displacement of all the atoms.

If the theory of second-order phase transitions is by now relatively well understood, the same cannot be said for first-order transitions. Actually, it may be misleading even to mention the two in the same breath, since the relevant questions are quite different in the two cases. Whereas in a second-order transition each phase is, as it were, acutely conscious of the other's existence, and macroscopic subvolumes of the system can fluctuate between the two phases, in a first-order transition the problems are associated precisely with the fact that the two phases in question are quite different and cannot easily communicate with one another. Thus, while there are no special problems connected with the thermodynamics of (say) a liquid in the supercooled region (no analogue of the critical behaviour near a second-order transition), the real question is how the more stable solid phase is ever 'nucleated' from the metastable liquid. From an experimental point of view, most first-order phase transitions seem to be very sensitive to factors such as contamination of the sample, external noise, and so on, and in some cases manifest behaviour which is highly irreproducible between one run and the next. This is not particularly surprising. Developing the analogy used earlier, one may think of a system about to undergo a first-order phase transition as like a billiard-ball which rolls in a hollow at the top of a hill and is buffeted by random blows of random magnitude in all directions; eventually, it will certainly be knocked over the edge of the hollow and will roll down the hill, but it is extremely difficult, if not impossible, to predict exactly when, or even how, this will happen. In the case of interest the 'random blows' are provided by the noise inherent in the system and its environment, and we do not always have a very good understanding of how this noise arises. It is produced by fluctuating forces whose individual microscopic origin is complicated or unknown, and which therefore appear to us to be random, like the unwanted 'noise' in a radio receiver. In particular, there is one type of noise which appears to be fairly ubiquitous, the $1/f$ noise (so called because the magnitude of the fluctuations at a given frequency f is roughly inversely proportional to the frequency); this is still very poorly understood as a general phenomenon, although in recent years some possible origins of it in specific systems have become much clearer. Thus, it is not too surprising that we should lack a detailed quantitative understanding of the transition from a metastable to a stable state.

Even given this uncertainty in detail, however, there are at first sight some general statements we ought to be able to make concerning first-order transitions. One of these follows from the fact that to make the transition from the metastable state to the thermodynamically stable one the system must pass through states of higher free energy than the original metastable state (just as our billiard-ball has to climb to the edge of the hollow before it can roll down the hill). We say that it has to 'overcome a free energy barrier'. Now, as we saw earlier, the condition necessary for a system (or part of it) to fluctuate into a state of higher free energy is that the cost in free energy should be comparable to, or not much larger than, the thermal energy at the temperature in question. If the cost in free energy is too large, greater than fifty times the thermal energy, say, then the process is so improbable that one can rule it out. Thus, if we know the height of the free energy barrier at a given temperature, we should be able to predict whether or not nucleation of the stable phase will ever occur (with any reasonable probability) at this temperature. Such a calculation can be done for many real-life first-order transitions; in some cases it gives good agreement with experiment, while in others it considerably underestimates the probability of transition. In the latter cases it is often plausible at least to attribute the discrepancy to extrinsic factors, such as contamination by 'dirt' in various forms, which can lead to an increased probability of nucleation. However, there are one or two cases in which this explanation seems to be ruled out—either because 'dirt' of the appropriate kind is known not to be present, or because one can show on rather general grounds that it could not lower the free energy barrier by the necessary amount. In particular, there is one spectacular case, involving the superfluid phases of liquid ^{3}He (on which, see below), in which the discrepancy between the predicted lifetime of the metastable state against nucleation—many times the lifetime of the universe—and the observed lifetime—less than five minutes—is so striking, and so clear-cut, that one is seriously forced to wonder whether the whole statistical-mechanical basis of our current picture of the nucleation process isn't in need of drastic revision.

In considering first-order phase transitions we have implicitly assumed that there is a single metastable phase, as well as (by definition!) a single stable one. An even more interesting class of systems is that in which there is a variety, perhaps an infinite variety, of metastable states. A well-known example is ordinary window glass, which is composed mainly of silicate molecules, which in turn are formed of silicon and oxygen atoms. It is possible, of course, to form from these molecules a regular crystalline array, and indeed, if the solid is formed under appropriate conditions, this is what we get. However, in the manufacture of ordinary glass, the system is cooled from the liquid phase too rapidly for the molecules to have time to fit together in the optimum way; instead, they end up all jumbled together, without any regular or repeating pattern—a so-called amorphous solid. One property of amorphous solids, or 'glasses', which distinguishes them from regular crystals is that not only are there very many metastable states, but that the degree of accessibility of the latter to one another varies enormously. To change the configuration of a few neighbouring molecules may be quite easy—that is, it may involve crossing only a relatively low free energy barrier—whereas to rearrange the whole system—for example, so as to form the crystalline array which is actually the thermodynamically stable state—may involve crossing a barrier that is so high that it will not happen in millions of years. Thus, those properties of the glass which depend on the relative motion of the molecules (such as specific heat and thermal conductivity) may show very complicated behaviour, quite unlike anything seen in crystalline solids; in particular, the apparent specific heat may be time-dependent, since the longer we leave the system, the larger the number of ways of sharing the heat that become available to it, as more and more improbable transition processes come into play. A simple glass like window glass is actually only a special case of a much wider set of systems with qualitatively similar features; other examples are solutions of long-chain polymer molecules (such as rubber) and a so-called spin glass, a system of atomic magnets interacting by forces which are random in magnitude and sign. All these systems have in common that the shapes of the interacting elements and/or the form of their interactions are sufficiently complicated that it is very difficult—usually impossible, in fact—for them to find the true minimum of the free energy, and so they have to be content with sliding from one metastable state to another. The static properties and, even more, the dynamics of such systems, are currently a topic of great interest.

Apart from their intrinsic interest, one important motivation for the study of such 'glassy' materials is the hope that it will lead to some insights into the physical behaviour of what is perhaps the most fascinating class of condensed-matter systems of all—namely, biological systems. It is remarkable that while we are often able to make quite detailed predictions of the properties of some newly synthesized crystalline compound, often under quite exotic conditions of temperature, pressure, and so on, we remain very largely ignorant of the precise physical mechanisms which allow the efficient functioning of the biomolecules composing our own bodies, even at room temperature. That is not to say, of course, that we do not know what atoms compose these molecules; we often know this very well indeed, and are able to make detailed and accurate three-dimensional models of the molecules in question. Also, in many cases we have quite a detailed knowledge of the chemistry of specific reaction processes. What we understand much less well in physical terms are Nature's design criteria, as it were, even at a very specific level. For example, a given enzyme molecule may catalyze a particular biochemical reaction, and we may be able to pinpoint with very high accuracy the exact region of the enzyme where the reaction takes place, and even to understand in some detail the chemistry of the processes involved. But the enzyme in question may be very large—much larger than at first sight seems necessary—and the removal of a small and apparently irrelevant group of atoms far away from the reaction site may completely spoil the enzyme action. Clearly there has to be some subtle physical mechanism involved which cannot be 'local' in nature, but in many cases we do not even begin to understand what that might be. Again, consider the similarities and differences between two so-called haemeprotein molecules, haemoglobin and myoglobin. Their biological roles, though related, are quite different: haemoglobin must capture oxygen in the lungs, transport it through the blood, and release it in the muscles, whereas myoglobin must store the oxygen in the muscles until it is needed. Regarded as physical structures, however, the two molecules are quite similar: to first appearances, haemoglobin looks very like four myoglobin molecules put together, and the differences between the detailed three-dimensional structures of the protein part do not look particularly impressive. Yet it is clear that they must in fact be very significant in allowing each molecule to fulfil its biological role.

What significant features of the physics do these differences affect? Could Nature have designed them differently? And how, historically speaking, did they come to evolve in the way they did? One reason why it is often thought that the study of glassy systems may be of some relevance to this kind of question is that it is very obvious indeed from the point of view of thermodynamics that functioning biological systems must be in a metastable state most, if not all, of the time, both at the level of the organism as a whole and even at the level of individual molecules. For example, a functioning protein is almost never found in a configuration corresponding to the lowest free energy available to it, or even anywhere near it. Indeed, the way in which biological systems generate complexity out of simplicity may seem at first sight to violate quite general conceptions of thermodynamics. Of course, there is no real violation once one takes into account external energy sources and so on; but it is clear that the sorts of physical concepts one may need to grasp the functioning of a complex biological system may be quite different from those familiar in the study of inert inorganic materials, and this remains a major challenge to condensed-matter physics.

Before leaving the subject of biological systems, we will amuse ourselves by considering one specific question which illustrates how seemingly unrelated areas of physics may unexpectedly make contact. Many biological molecules, including many of the amino acids which are components of proteins, have a 'handedness'—that is, they exist in two inequivalent forms which are mirror images of one another, like a right- and a left-handed glove. If we attempt to synthesize these molecules in the laboratory from their basic molecular components, we produce right- and left-handed molecules in approximately equal numbers. In nature, however, only one species occurs—the one conventionally regarded as the 'left-handed' variety. Is this an accident? That only left-handed *or* only right-handed molecules should occur in the world we know is perhaps not so mysterious, since if at any time in the course of molecular evolution both species were present, they might well have been in competition for the materials needed to reproduce themselves, and a small preponderance of one or the other, perhaps due to some random fluctuation, could easily have led to the preponderant species completely wiping out the other one. But is it an accident that the left-handed variety has in fact prevailed? Or was the competition subtly biased in its favour? To the extent that we believe that the laws of nature are identical in a right-handed and a left-handed system of coordinates, of course, there can be no such bias. But we saw in Chapter 2 that while this is true for the strong, electromagnetic, and gravitational interactions, it is not true for the weak interaction, which indeed has a preferred handedness. Moreover, while it was proposed many years ago that the handedness of the electrons emitted in beta-decay might somehow bias the evolution process, a new ingredient was added to the puzzle in the last ten years or so by the discovery of the parity-violating interaction, albeit tiny, between the electron and the nucleon and between electrons themselves which is associated with exchange of the Z^0 particle. Whether this tiny intrinsic bias is adequate to account for the observed dominance of left over right in contemporary biological systems (and if so how) is still a matter of lively debate.

There is one rather general way in which the frontier areas of condensed-matter physics being explored today differ from the traditional subject-matter of the discipline. In both cases, one first decides what one thinks are the physically most relevant features of the real-life system, then embodies them in some kind of idealized model, and then tries to 'solve' the model—that is, as far as possible, to make rigorous deductions from it. In traditional solid-state physics the most basic models—for example, the ball-and-spring model of a crystalline lattice or the theory of electrons moving freely in a metal—were usually exactly soluble by analytical mathematical techniques (that is, the solutions could be expressed in terms of familiar and tabulated mathematical functions such as sine, cosine, exponential, and so on). However, anything more complicated (such as the model of interacting electrons and atoms mentioned above) usually could not be solved analytically, and as a result one had to make so-called 'physical approximations'. In most of today's frontier areas, the situation is far worse, in that even the most basic models of, for example, a glass or spin glass are not usually soluble in closed analytical form.

At this point, however, modern computers have come to the rescue; even when there are no known ways of solving the mathematical problem analytically, it may be possible to generate a numerical solution, and provided one chooses the questions to be asked carefully, this can be just as informative. In fact, some current models of condensed-matter systems lend themselves very well to solution by so-called special-purpose computers, which are designed and built with a particular kind of problem in mind. Thus, one is often in a position to check the mathematical guesses which one would formerly have had to justify by 'physical arguments'. Of course, it may happen that as a result of seeing the computer-generated output of one's model, one comes to the conclusion that the model itself was not an appropriate description of the real-life physical system, but at least one possible source of uncertainty has been eliminated! One very important general conclusion that can be reached by so-called computer simulation is that once one leaves the class of analytically soluble models, even very simple models can produce extremely complicated behaviour, so complicated that any attempt at even qualitative prediction of the outcome usually fails. Moreover, it can be verified experimentally that at least some real physical systems are apparently well described by such models. For example, consider the simple system formed by a weight on the end of a string suspended from a fixed support. If this system moves only in the earth's gravitational field, it behaves as a classical (two-dimensional) pendulum, and in general simply traces out the same (elliptical) orbit over and over again (at least to the extent that its displacement from equilibrium is small). However, we have only to introduce a slightly more complicated force on it (for example, as in a currently available toy, by replacing the weight by a small bar magnet and placing two more magnets on the surface below) for it to move in a wildly unpredictable way, which is sensitive to the smallest variation in how it was started—a so-called chaotic orbit. If a macroscopic mechanical system as simple as this can behave in so complicated and unpredictable a manner, there seems no reason to doubt that the much more complicated models we have built of systems like glasses, spin glasses, and biomolecules will in principle also allow this kind of chaotic behaviour. At the same time it is a remarkable fact that, with some provisos, our familiar thermodynamics works for real-life glasses and spin glasses, and that biomolecular systems show quite remarkably reproducible, and even adaptive, behaviour. So while there are indeed *some* macroscopic systems whose chaos precludes accurate prediction (ask any weather-forecaster!), in many of the more interesting condensed-matter systems the effects seem to average out on a macroscopic scale. Just how this happens, and moreover whether the concept of chaos can legitimately be taken over into quantum mechanics, are questions of great current interest.

The problems I have discussed so far have to do with systems which occur naturally in the world around us, or which, even if artificially produced, are qualitatively similar to naturally occurring ones. However, one of the most exciting aspects of condensed-matter physics is the creation of qualitatively new types of system which Nature herself has not got around to producing. For example, it is now possible to design a new type of solid consisting of superimposed wafers of different materials each only a few atomic layers thick. Such so-called 'quantum well structures' allow the electrons in the solid to show behaviour that is in some ways qualitatively different from that in ordinary solids and will very probably usher in a whole new generation of ultra-small electronic devices. Again, in the man-made device known as a laser, one arranges things so as to produce a very large number of photons occupying exactly the same state, in a way which, as far as we know, does not occur in nature (at least in the optical region of the spectrum), and the resulting radiation has many unique features, some of which have been put to practical use.

Apart from these specific examples, however, there is one whole area of condensed-matter physics which can be described as a 'new frontier'—namely, the physics of very low temperatures, cryogenics. What distinguishes this area is that, if any of the ideas currently orthodox in cosmology are even approximately right (see Chapter 3), the present temperature of the cosmic blackbody radiation, about 3 degrees absolute, is the lowest that the universe has ever known. It follows that when we cool matter below this temperature, we are exploring physics which Nature herself has never explored. Indeed, to the extent that we exclude the possibility that on some distant planet other conscious beings are doing their own low-temperature physics, the phenomena unique to this region have never previously existed in the history of the cosmos.

Why should one expect the low-temperature field to produce anything qualitatively different from what goes on at room temperature? The first point to make is that nowadays low temperatures really are 'low': while the temperature of the cosmic blackbody radiation is about one-hundredth of room temperature, we can now cool matter in bulk to something like one-ten-millionth of room temperature. Now whether a particular type of force or interaction between particles can have an appreciable effect depends, crudely speaking, on whether the strength of that interaction is large or small compared with the characteristic thermal energy, which is proportional to absolute temperature. Thus, at very low temperatures we can investigate the effects of very small and subtle interactions which are of the order of one-ten-millionth of those which are observable at room temperature. In particular, co-operative behaviour which would be totally obscured by thermal noise under ordinary conditions may show up quite easily, and even spectacularly, in a low-temperature experiment.

Of the various qualitatively new phenomena occurring in the low-temperature region (not necessarily only below 3 degrees), probably the most fascinating is the complex of effects which go under the name of 'superconductivity' when they occur in an electrically charged system such as the electrons in a metal, and 'superfluidity' when they occur in a neutral system such as an insulating liquid. The behaviour of a superconducting metal is qualitatively quite different from that of a normal metal, in that it appears to conduct electricity with zero resistance (hence the name), it expels any magnetic flux applied to it, sometimes with dramatic effects, and so on. Similarly, a liquid which goes superfluid, such as the common isotopic form of helium, can flow through tiny capillaries without apparent friction, can climb up over the rim of a vessel containing it and thereby gradually empty the vessel, and so on.

According to our current understanding, these bizarre phenomena are dramatic manifestations of intrinsically quantum-mechanical collective behaviour. To sketch the general picture, let us focus on the common isotope of helium (^{4}He). Such an atom has zero total spin, and therefore, according to the spin-statistics theorem mentioned in Chapter 2, should obey Bose–Einstein statistics. A remarkable prediction of statistical mechanics for particles obeying such statistics is that when a large number of such particles are placed in a restricted volume and are cooled down below a certain temperature (which for helium turns out to be about 3 degrees), a phenomenon called 'Bose condensation' should take place: namely, a finite fraction of all the atoms should begin to occupy a *single* quantum-mechanical state, the fraction increasing to unity as the temperature decreases to zero. The atoms in this special state become locked together in their motion, like soldiers in a well-drilled army, and can no longer behave independently. Thus, for example, if the liquid flows through a narrow capillary, the processes of scattering of individual atoms by roughness in the walls and so on, which would produce so strong a viscous drag on any normal liquid as to effectively prevent it from flowing at all, are now quite ineffective, since either all atoms must be scattered or none. The quantum-mechanical nature of the behaviour also has quite spectacular consequences: for example, if the liquid is placed in a doughnut-shaped container, the wave which in quantum mechanics is associated with a particle must 'fit into' the container—that is, there must be an integral number of wavelengths as we go once around (compare the discussion in Chapter 1 of the electron wave in the atom). This is also true in a normal liquid, but the effects there are quite impossible to see, because each atom tends to occupy a different state with a different associated number of wavelengths. In superfluid helium, on the other hand, the fact that all (condensed) atoms are in the same state makes the effect spectacular: for example, the liquid can rotate only at certain special angular velocities and, if the container is rotated, may therefore get 'left behind'. This is a special case of a much more general feature: effects which in a normal system would be totally obscured by thermal noise—that is, the fact that every atom is doing something different—are often quite easy to see in a superfluid system.

A similar general picture is believed to apply to superconducting metals, except that the 'particles' which undergo Bose condensation (and are therefore required to be bosons—that is, to have integral spin) are not individual electrons—which in any case have spin $\frac{1}{2}$ and are therefore fermions, not bosons—but rather *pairs* of electrons which form in the metal, rather as some atoms form diatomic molecules. Something rather similar happens in a liquid composed of the rare isotope of helium (^{3}He), where the basic entities (the atoms) are also fermions. Here, however, there is a further interesting feature—namely, that the fermion pairs which undergo Bose condensation have a non-trivial, and variable, *internal* structure, which by the nature of the Bose condensation process must be the same for all pairs. As a result, it may be possible to use ^{3}He to search for various kinds of effects which are far too tiny to be seen at the level of an individual molecule: the superfluid condensation 'amplifies' them to a possibly observable level. It is likely that the phenomenon of superfluidity also occurs in other systems of fermions—for example, in the neutrons in a neutron star, although the evidence here is necessarily more circumstantial.

The phenomena of superconductivity and superfluidity were not, and probably could not reasonably have been, predicted ahead of their experimental discovery in various metals and in ^{4}He respectively; indeed, for many years they constituted a major mystery. What they show beyond any doubt is that interactions between a very large number of 'elementary' particles can give rise to macroscopic effects which are qualitatively totally unexpected, effects which could never have been deduced by studying small groups of particles. Thus, however true it may be that these phenomena are 'merely' consequences of well-understood electromagnetic interactions between the elementary particles involved, the 'merely' begs some important questions. Almost certainly there are other qualitatively new phenomena waiting to be discovered in condensed-matter systems; indeed, it is entirely possible that some of them have already been seen in existing experiments, but have been dismissed as due to experimental error (as in fact happened with at least one important effect in superconductivity). To predict such qualitatively new collective effects in advance—a task in which, it should be said, with very few exceptions such as the laser, it has not so far been very successful—or, more modestly, to build explanatory models for them once they are discovered experimentally, will surely remain a major goal of condensed-matter physics and a major part of its attraction for its practitioners.

Quite apart from specific systems and phenomena, such as those mentioned in this chapter, I believe that there is one very general and fundamental aspect of condensed-matter physics which is likely to attract increasing attention in the future—namely, the question of how we 'interface' the specific system or sample we wish to study with its environment, including the instruments we use to prepare it in a given state and to make measurements on it. Suppose, for example, that we wish to measure the electrical conductivity of a given piece of metal (the 'system') at low temperatures. To do this, we need to attach to it electrical leads, and to apply to these leads either a known, fixed current or a fixed voltage; moreover, we need to put it in contact with a thermal 'bath' (for example, a literal bath of liquid helium) which will maintain its temperature at the required level. Within the framework of classical physics there is no particular problem, either experimental or theoretical; it is entirely consistent to suppose that we can indeed apply (say) a fixed current, which we can measure with an ammeter, if desired, without disturbing the system or affecting its response; and, moreover, that we can take the effect of the thermal bath into account by assuming that it exerts on the system microscopic forces which are random but well-defined in their statistical properties. Even when the system itself has to be described by quantum mechanics, it has traditionally been assumed that this essentially classical framework for handling its interactions with its environment remains valid; and indeed, all the standard results of quantum statistical mechanics—that is, statistical mechanics as applied to quantum systems—are implicitly based on it.

Over the last ten years or so, however, a quiet revolution has been taking place in experimental condensed-matter physics, a revolution whose ultimate implications are as yet unclear, but are likely to be profound, I believe. With the aid of advances in cryogenics, microfabrication, and noise isolation techniques which would probably have seemed out of the question only a few years ago, we are now rapidly approaching the point—indeed, in some contexts have already reached it—at which quantum effects in the system can be amplified to the point where they, as it were, cross the line separating it from its environment. The best-known example of such a situation concerns the question of the extent to which the quantum-mechanical wave which represents an electron travelling through a metal can preserve its phase. It turns out that, crudely speaking, collisions with static impurities preserve the phase, but collisions with phonons tend to change it in an unpredictable way, thereby washing out the characteristically quantum-mechanical interference effects. In traditional treatments of the electrical resistance of a metal, it has almost always been assumed that many such phase-destroying collisions will take place during the passage of the electron through the sample. However, experiments have recently been done with small enough samples and at low enough temperatures that not even one such collision occurs, and the results make it clear that the whole concept of resistance needs to be rethought under these circumstances; in particular, the precise details of the way the leads are attached, the current controlled, and so on become of significance in a way quite at odds with the traditional conceptual framework. A similar situation has been reached, or will probably be reached in the near future, with regard to a number of other problems in condensed-matter physics which involve the 'amplification' of quantum effects to the macroscopic level, whether in terms of geometrical scale, number of particles involved, or other variables.

In this situation, it is a real question whether our traditional ideas about the effect on a test 'system' from its 'environment' will continue to work, or whether the whole idea of separating the two conceptually will break down. Such an eventuality might well force us to revise drastically many of our prejudices about the behaviour of condensed-matter systems. It could even be that we shall have to start worrying seriously in this context about paradoxical aspects of quantum mechanics of the sort which are spectacularly displayed by Bell's theorem (see next chapter). In other words, we may run up against a 'quantum preparation problem' which is every bit as severe as the 'quantum measurement problem' to be discussed in the next chapter. But, after all, this is only a small cloud on the horizon . . .

5
Skeletons in the cupboard

This chapter should be prefaced with the intellectual equivalent of a government health warning. With the possible exception of some of the remarks at the beginning of the last chapter, where some might disagree with the emphasis, I believe that the picture of the current state of physics which I have given would seem broadly reasonable to most contemporary practising physicists. By contrast, some of the views to be explored in this chapter, particularly towards the end, would probably be characterized by the more charitable of my colleagues as heterodox, and by the less charitable quite possibly as crackpot. Indeed, many may well feel that excessive concern with these problems could irreparably damage one's intellectual health. Nevertheless, I make no apology for presenting them here: were there no issues in physics which continue to rouse strong feelings and probe people's deepest intellectual commitments, one could be pretty sure that the subject was withering at the roots.

In this chapter I shall discuss, in progressively greater detail, three superficially unrelated issues: the so-called anthropic principle, the 'arrow of time', and the quantum measurement paradox. Although these are associated historically with quite different areas of physics—cosmology, statistical mechanics, and quantum mechanics, respectively—they have much in common. In each case it is probably fair to say that the majority of physicists feel that there is simply no problem to be discussed, whereas a minority insist not only on the problem's existence, but also on its urgency. In no case can the issue be settled by experiment, at least within the currently reigning conceptual framework—a feature which, as noted in Chapter 1, leads many to dismiss it as 'merely philosophical'. And in each case, as one probes more deeply, one eventually runs up against a more general question, namely: In the last analysis, can a satisfactory description of the physical world fail to take explicit account of the fact that it is itself formulated by and for human beings?

The anthropic principle

The anthropic principle in cosmology has a long and venerable history—in fact it goes back effectively to long before the birth of physics as we know it—and it is possible to distinguish a number of different subspecies; I shall not attempt to go into these fine distinctions here.[1] Most modern versions start from two general observations. The first is that, in the current formulation of particle physics and cosmology, there are a large number of constants which are not determined by the theory itself, but have to be put in 'by hand'. For example, in current particle theory the ratio of the electron and proton masses is not fixed by the theory; nor is the basic unit of electric charge, e (which is dimensionless when expressed in units involving the fundamental constants h and c); nor is the number of different quark flavours, and so on. Even more fundamentally, neither particle physics nor cosmology is able to tell us why we live in a world with one time and three space dimensions, or why space should be (apparently) isotropic, and so on. The second observation is that the physical conditions necessary for the occurrence of life as we know it—still more for the development of life to the human stage—are extremely stringent. For example, the biochemical reactions essential to life depend extremely sensitively on the energies of the molecular states involved, which in turn are very sensitive to the exact value of the electron mass and charge, among other things; were the electron charge only very slightly different from what it in fact is, the biochemistry we know could not exist. Again, the development of life apparently requires not only the right chemical conditions but just the right distribution of incident radiation: were the ratio of electromagnetic and gravitational interaction constants only slightly different from what it actually is, our sun would be unable to provide the correct mix. In the context of recent grand unified theories, which require the proton to be unstable, with a lifetime of the order of 10^{32} years, one observes that a lifetime a few orders of magnitude shorter than this would have us all dying of radiation sickness. Again, in a world which had other than three space dimensions the gravitational force would not follow an inverse-square law, stars as we know them could not exist, hence no planets . . . and so on. The list could be multiplied endlessly, and it is easy to draw the conclusion that for any kind of conscious beings to exist at all, the basic constants of nature have to be exactly what they are, or at least extremely close to it. The anthropic principle then turns this statement around and says, in effect, that the reason the fundamental constants have the values they do is because otherwise we would not be here to wonder about them.

Before exploring the logical status of this 'explanation', it is perhaps as well to note that it is easy to overstate both the supports on which it claims to rest. It is true that current particle theory has a large number of undetermined constants (seventeen at the last count), but some progress has been made over the last few years in reducing the number—for example, in the unified electroweak theory—and it seems a reasonable hope, and certainly one shared by many theorists in the area, that the development of further theories of the grand unified type will reduce it still more. There is even some hope that the so-called superstring theories currently under intensive study will eventually determine that the only fully consistent version of quantum field theory inevitably gives rise to a world which, at the energy scale we encounter in everyday life, not only contains the particles we observe, but necessarily has one time and three space dimensions. (Of course, it would still be possible to ask why nature has to be described by a quantum field theory at all, but that is a question at a rather different level.) On the other hand, the fact that life *as we know it* seems to depend very critically on the fundamental constants having the values they do may not be as significant as it seems. As has been repeatedly demonstrated in the area of condensed-matter physics and elsewhere, Nature is a great deal cleverer than we are, and it is not so difficult to imagine that were the proton lifetime indeed so short as to pose a danger to the living systems we know, for example, biology would have evolved in a slightly different way to cope with the problem. A further point is that while some of the undetermined inputs into current theories, such as the dimensionality of space, can take only discrete values, others, such as particle masses and charges, can take a continuous range of values (at least within the framework of current conceptions). It is not clear that very much has been gained by an 'explanation' which merely pins the values down within a finite range, however narrow.

Leaving aside these points of detail, is the anthropic principle Humpty-Dumpty logic? The answer must surely depend on the metaphysical framework within which one views it. Within the framework of a teleological point of view—that is, one in which the universe is seen as designed, or as having evolved, specifically in order to permit the existence of man—such a principle seems completely natural. However, modern physics, at least as currently practised, rejects teleological arguments, and even those physicists who, as a matter of religious belief or for other reasons, regard the universe as purpose-designed generally feel it incumbent on them in their everyday work to seek other kinds of explanation to supplement the teleological one. If, therefore, one stays within the currently established metaphysical framework of physics, it seems to me that the plausibility or otherwise of the anthropic principle as an 'explanation' depends crucially on what exactly it claims to explain. If the claim is that there is, as it were, only one universe, in which the constants of nature are fixed for all times and places, and that the principle explains why they have the values they do, then I personally believe that the word 'explain' is being used in a sense so different from its normal usage in the currently established framework as to be, within this framework, quite unintelligible.

However, recent advances on the cosmology–particle physics interface have made at least thinkable a scenario in which the principle might conceivably play an intelligible rôle. Suppose that we accept the hot big bang hypothesis, and believe that in its very early stages the universe was in a state in which various symmetries were unbroken. As its evolution progressed, some of these symmetries might have been broken; and if the breaking had occurred independently at points in space–time which were causally disconnected—that is, had space-like separation—then the nature of the broken-symmetry phase could be different in different regions. (An analogy would be the case of a liquid which can crystallize in more than one crystal structure: if the ends of the vessel are cooled rapidly while the middle remains hot, different structures can form at the two ends.) It is conceivable that there could be an infinity of possible broken-symmetry phases, each with its own characteristic values of the particle masses, coupling constants, and so on (compare the effectively infinite number of different configurations in which a glass can freeze); and, given an initial theory with sufficiently high dimensionality and sufficiently complicated structure, it is even not unthinkable that the effective dimensionality, at 'laboratory' energies, should be different in different regions of the universe which cannot communicate with one another. To be sure, this scenario is speculative in the extreme. If there is any truth in it, however, there might be a role for a modified anthropic principle: the reason or, more accurately, part of the reason, that *in the special part of the universe accessible to us* the fundamental constants have the values they do is that in other parts of the universe, where the constants are different, we simply could not have evolved. Of course, even this more modest claim would raise further questions—for example, is there anything which in any way intrinsically distinguishes 'our' part of the universe *other* than the fact that we have evolved in it?—but I will not pursue the issue further. The only reason I have included this brief and of necessity superficial discussion of the anthropic principle at all is to illustrate the point that at the very heart of those areas of physics which are often regarded as most fundamental, there lie questions where the very concept of what constitutes an 'explanation' is itself still a matter of lively debate.

The 'arrow of time'

It is almost too obvious to be worth stating that the world around us seems to exhibit a marked asymmetry with respect to the 'direction' of time; very many sequences of events can occur in one order, but not in the reverse order. To make this a bit more precise, let us imagine ourselves viewing a film of some scenes in everyday life which has been put into a projector backwards. It does not take us long to realize this. For example, if we are shown a bouncing ball, the height of the bounces increases with time and the ball eventually makes for the hand of a child; a pool of liquid on a table contracts and runs back into an overturned teacup, which then rights itself; a poker, cold at first, gradually begins to glow without any obvious source of heat; and so on. As the last example shows, it is not necessary for any human agents to be explicitly involved for us to recognize that something is wrong—that we are seeing a 'time-reversed' version of the original. Thus, quite apart from our subjective sense of 'past' and 'future', it looks as if Nature herself 'cares about' the sense (direction) of time.

Why is this simple observation in any way problematic? The reason is that, at first sight at least, it conflicts with the symmetry of the laws of physics at the microscopic level with respect to a reversal of the sense of time. Consider, first, classical Newtonian mechanics. Newton's first and third laws clearly do not refer to the sense of time, and would have identical forms in a time-reversed system. As to his second law, the acceleration which appears in it is the *second* derivative of position with time; so if we reverse the sense of time, the velocity (and momentum) is reversed, but the acceleration is unchanged, and thus the second law also is the same in the time-reversed system. Consequently, we conclude that in Newtonian mechanics, to every possible sequence of events there corresponds another, equally possible sequence which is its 'time-reverse'—that is, the sequence is traced in the reverse order. For example, if we are shown a speeded-up film of the motion of the planets of a distant star, it is impossible to tell simply by inspecting the film whether or not it is being run backwards. Again, if we see a film of billiard-balls moving and colliding for a short time on the surface of a billiard-table (without being struck by a cue or falling into pockets), then, to the extent that friction with the cloth and inelasticity in the collisions can be neglected, it is again impossible to tell whether or not the film is being shown the right way round. In both these cases we are dealing with macroscopic systems in which dissipative effects are to a first approximation negligible; however, if we take such effects into account, the time-reversal symmetry is spoiled.

So can we invoke dissipation to introduce a fundamental asymmetry? Were dissipation a completely new phenomenon quite unconnected with Newton's laws, that might be plausible. But in the modern molecular view of matter, dissipation means nothing but the transfer of energy from the macroscopic motion into the random motion of molecules; for example, when two billiard-balls collide, some of their kinetic energy is lost, but this energy reappears as heat—that is, the random motion of the molecules composing the balls. At the molecular level of description, Newton's laws continue to apply (or so we believe), and any possible sequence of events in a system of molecules has an equally possible time-reversed counterpart.[2] Indeed, were we able to take a film of the motion of the molecules of (say) a gas in a state of thermal equilibrium, there would again be no way of telling whether or not it was being run backwards; and, while a process in which heat was converted, in a collision, into increased kinetic energy of the billiard-balls would certainly look very odd, it would in no way violate Newton's equations of motion. So, at the classical level at least, there is no microscopic asymmetry which we can use to explain the observed macroscopic one.

Does quantum mechanics help? Unfortunately not. While the standard textbook formulation *looks* at first sight as if it is asymmetric with respect to time (the equation governing the evolution of the wave function in non-relativistic quantum mechanics—Schrödinger's equation—is first-order in time, not second-order like Newton's second law), the appearance is misleading. In fact, with the exception of one puzzling phenomenon in particle physics which I will mention below, in quantum mechanics, as in classical mechanics, any process which is allowed has a corresponding time-reversed process which is also allowed. So we are back to the same difficulty.

Perhaps it is as well at this point to dispose of an obvious objection which may have occurred to some readers. Is it not the case that in electromagnetism, at least, the two directions of time are not equivalent? For example, a positively charged particle in a magnetic field, when viewed along the direction of the field, will appear to circle clockwise; and the time-reversed motion—that is, a circle executed anticlockwise—is not possible. However, the magnetic field is itself due to circulatory currents (or the equivalent), so when we reverse the sense of time, and hence the direction of the currents, we must for consistency also reverse the magnetic field—and the anticlockwise motion *is* then indeed possible, in fact mandatory! Again, the violation of time symmetry is only apparent.

Before we go on to discuss the way in which statistical mechanics attempts to account for the observed asymmetry, one related issue needs to be mentioned. As I emphasized in Chapter 1, Newton's second law has the form of a second-order differential equation with respect to time, and if we know the relevant forces, a complete determination of the solution requires two independent pieces of information. In many contexts it seems natural to take as these two pieces of information the data we have on the *initial* state of the system—say, the initial position and velocity. But Newtonian mechanics itself in no way forces this choice on us, and we could equally well take as our data the *final* position and velocity, and then calculate the previous motion from them. So there is no support in the formalism of Newtonian mechanics for the idea that 'the past causes the future', rather than vice versa. Again, quantum mechanics provides no help here, for although the standard textbook formalism assumes that the initial state of the system is known, and that Schrödinger's equation (or its relativistic generalization) then 'determines' the subsequent behaviour, it is easy to show that there is a fully equivalent formalism in which it is the *final* state of the system which is known, and which 'determines' the earlier behaviour. So the idea that event A can cause event B only if it precedes it is an *interpretation* of the formalism of physics, not something which follows necessarily from it. (It *is* a consequence of special relativity that two events, A and B, can be causally related only if they can be connected by a light or a slower-than-light signal; but special relativity says nothing about the *direction* of the causal connection. The surprisingly widespread notion that certain experiments in particle physics have 'proved that causality propagates forwards in time' is simply a misconception; on close examination it turns out that this assumption is already built into the interpretation of the experiments in question.) Note that this question is of a quite different nature from that of why certain kinds of processes appear to go only one way in time; the latter refers to an experimentally verifiable fact, whereas the former is a question about the interpretation we put on experiments. Indeed, one might well argue that the whole notion that in physics or cosmology we can or should 'explain' the present by reference to the past may be an illegitimate residue of anthropomorphic thinking. Why not 'explain' the past by reference to the present?

Let's return to our original problem, and consider a specific example. Suppose we take a container which can be divided in half by a movable partition, put the partition in place, and fill one half—the left half, say—of the container with gas (Figure 5.1). So long as the partition remains in place, the gas is confined to the left half. Then at some subsequent time we remove the partition. In a very short time the gas will expand to fill the whole container, and however frequently we observe it thereafter, we

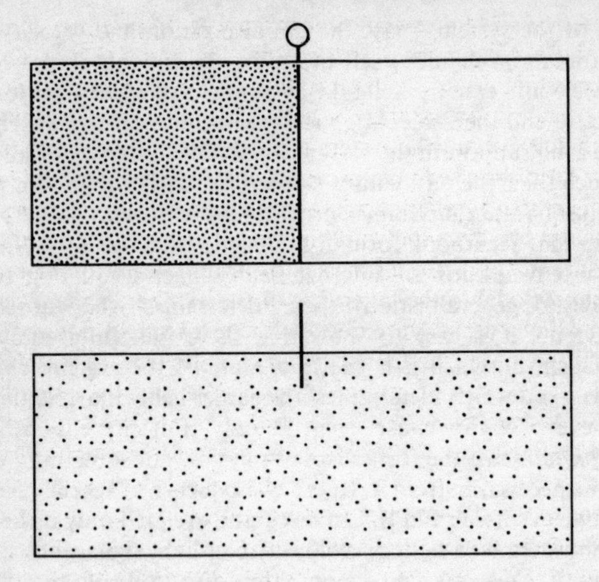

Figure 5.1 A gas confined to a small volume which is part of a larger one (*top*); the same gas after expanding into the whole large volume

will never find it back in its original state—that is, confined to the left-hand side of the vessel (unless, of course, we take a hand in the matter ourselves—for example, by inserting a piston and re-compressing the gas). Let's denote by A the state of the gas immediately after the partition was removed (but before it has had time to expand) and call B the final state in which it fills the whole volume; then it is clear that the state A can and does develop into state B, but not vice versa, so the process A→B is irreversible.

At first sight, statistical mechanics seems to offer a very straightforward and natural account of why this should be so. The point is that the description of the gas as being in state A or state B is of course a *macroscopic* ('thermodynamic') description, and in each case is compatible with a vast number of different possibilities for the positions and velocities of the individual molecules. However, it is clear that there are many more possibilities available when the gas is in state B than when it is in A: the range allowed for the position of each molecule is twice as great, so for n distinguishable molecules, the total number of possibilities is 2^n times what it is in state A. In practice, n might be of the order of, say, 10^{20}, so that the number 2^n is almost inconceivably large. We can say that the 'degree of disorder' is much greater in state B than in state A; or, alternatively, that our 'information' about the system is much less in the former (since the system's observed macroscopic behaviour is compatible with a much greater number of microscopic possibilities). It turns out that the thermodynamic variable we call entropy is nothing but a measure of this disorder or lack of information (to be precise, it is proportional to the logarithm of the number of microscopic states which are compatible with our knowledge of the macroscopic properties of the system). The fact that the process A→B is irreversible is then just a special case of the second law of thermodynamics, which states that the entropy of an isolated system may increase with time, or at best stay constant, but can never decrease; or, to put it differently, that as long as we do not interfere with the system, our information about its microscopic state cannot increase.

What is the basis for this assertion? It is simply a special case of the rather general statement that less probable situations can evolve into more probable ones, but not vice versa. Consider the analogy of shuffling a pack of cards. Suppose we start with a pack in which the top thirteen cards are all spades (not necessarily in any particular order), the next thirteen are hearts, and so on. It is clear that shuffling the pack in a random manner will rapidly destroy this feature; moreover, if we start with a pack prepared in no particular way, the chances of arriving through the shuffling process at a situation where the four suits are separated as described is totally negligible. The reason is simple: if the shuffling is a truly random process, then each particular sequence of the individual cards (for example, three of hearts, king of clubs, and so on) is equally likely to occur, and the number of sequences which have the special property in question, though very large compared to one, is very small indeed compared with the total number of possible sequences. Just so, the probability of a set of gas molecules which start in state B evolving, by the random process of collisions with one another and with the walls, into one of the very few microscopic states (1 in 2^n) which correspond to state A is in practice totally negligible.

Incidentally, at this stage we can see that there is a relation between the question of irreversibility and the question of why we tend to think of the past as 'causing' the present, rather than vice versa. If we consider systems in which irreversible processes occur (as distinct, for example, from the motion of the planets), then there is a 'many-to-one' correspondence: many initial thermodynamic states A can lead to the same final state B, but not vice versa. Thus, given a particular final state B, it seems natural to ask from which of the possible initial states A it evolved, whereas no corresponding 'time-reversed' question can be sensibly asked. Whether this observation is in any sense a justification of our everyday notions of causality is of course a much more subtle question, which I shall not attempt to address here.

So have we explained the asymmetry of thermodynamic processes with respect to time? Unfortunately not. If we look back at the above argument, we will see that *as long as no human agency is explicitly or implicitly involved*, it is actually completely symmetric with respect to the sense of time. Let us suppose that we know with certainty that the shuffling of our pack of cards over a given period of time is indeed random—it is being done, for example, by a machine programmed with the help of a so-called random number generator. If at some point during that time we inspect the pack and find (to our great surprise!) that the four suits are indeed separated in the way described, then we can conclude with high confidence not only that it will be more disordered in the future, as the shuffling proceeds, but that it was also more disordered in the past. (To see this, it is sufficient to observe that the 'inverse' of any given shuffling operation is itself a shuffling operation.) Similarly, if our gas is indeed isolated, and if by some extraordinary fluke we observe it at a given moment to be in a state A (that is, with all molecules in the left-hand half of the container), we could infer not only that it will rapidly tend to state B, but that it was in state B in the recent past. But of course in real life we would be unlikely to draw these conclusions: if we came across a pack of cards lying on a table and found that the suits were separated, we would not normally surmise that this was the result, by some extraordinary statistical fluke, of some random shuffling process, but would rather draw the common-sense conclusion that someone had deliberately ordered the cards this way. Similarly, if we were presented with a container in which the gas molecules were (for the microsecond or so for which we were allowed to observe it!) all on the left-hand side, the common-sense hypothesis would be that someone had prepared it in the way described above and had only just removed the partition. In other words, we tend to believe, at least in simple laboratory situations, that any configuration we come across which is very improbable in the statistical sense must be a result of conscious human intervention in the recent past. So it seems that the reason why entropy, or disorder, appears always to increase with time is simply that we are able, as human agents, to prepare 'very improbable' situations and *then* leave them to themselves; that is, we can in effect determine the *initial conditions* for the behaviour of the system. By contrast, we cannot determine the *final* conditions (this impossible task is sometimes called 'retroparing' the system in a given state). This statement is just a generalization of the observation made in Chapter 1 regarding classical particle mechanics—namely, that at least in simple laboratory situations, we can determine the initial values of the position and velocity but not the final ones. So it would seem that the arrow of time which appears in the apparently impersonal subject of thermodynamics is intimately related to what we, as human agents, can or cannot do.

As a matter of fact, there is a great deal more to the problem than the above discussion would indicate. In the first place, it is clear that there are many processes in nature in which no human agency could have been involved in the 'preparation' of the system, yet the second law of thermodynamics still appears to hold—for example, in geophysics and astronomy. Indeed, in general discussions of the subject it is conventional to distinguish (at least) five different arrows of time. One is the 'thermodynamic' arrow, which we have already encountered, which is determined by the increase of entropy. A second, which I have argued is intimately connected with the first in at least some contexts, is the 'human', or psychological, arrow of time, determined by the fact that we can (or think we can!) remember the past and affect the future, but not vice versa. (Presumably this is connected with the fact that as biological systems we differentiate in a time-asymmetric way, so it is natural to assume that there is no 'biological' arrow which might be different from the psychological one. However, at least within the context of physics, this is not something of which we can claim much understanding at present.) A third arrow is the 'cosmological' one, determined by the fact that the universe is currently expanding rather than contracting. A fourth arrow is the so-called electromagnetic one, and this requires a little explanation. The equations which describe the interactions between a charged body and the electromagnetic field are symmetric with respect to the sense of time. In particular, when a charged body accelerates, there are two kinds of solution for the behaviour of the electromagnetic field. The one conventionally adopted is the so-called retarded solution, in which an electromagnetic wave propagates out from the accelerated charge—that is, the charge radiates energy 'to infinity'. However, there is an equally good solution, the so-called advanced solution, in which the electromagnetic wave propagates *in* 'from infinity' towards the charge, transferring energy to it. The equations themselves are compatible with either solution, or with a combination of the two; however, while in the case of the retarded solution the only initial conditions we need to specify concern the motion of the charge itself, for the advanced solution we would need to specify the electromagnetic fields on a large sphere surrounding the charge, and they would have to be exactly right. (The situation is similar to that of a pond in which waves are created by the motion of a vertical piston and radiate outwards: it is possible in principle to reverse the motion and get the waves to propagate inwards and transfer energy to the piston, but to do so we would have to start the waves out exactly right at each point on the bank of the pond.) Thus, the reason often given, or implied, for rejecting the advanced solution is that we do not believe that those particular initial conditions could happen 'by accident'. But once again, this clearly brings us up against the question of why we are happy to consider certain configurations as implausible when they are initial conditions, but not when they are final conditions. The situation here is clearly reminiscent of that encountered in the context of the thermodynamic arrow.

Finally, for completeness I should mention the fifth arrow, which is associated with a class of rare events in particle physics—the so-called CP-violating decays of certain mesons (see Chapter 2)—and probably with other very small effects at the particle level. This arrow is rather different from the others, in that the relevant equations are themselves explicitly asymmetric with respect to the sense of time, rather than the asymmetry being imposed by our choice of solution. At the moment the origins of this asymmetry are very poorly understood, and the general consensus is that it probably does not have anything much to do with the other arrows; but it is too early to be completely sure of this. (Compare the rather similar situation, mentioned in Chapter 4, with regard to parity violation at the subatomic and the biological levels.)

Regarding the first four arrows, a point of view which seems free from obvious internal inconsistencies at least is that one can take as fundamental the cosmological arrow; that the electromagnetic arrow is then determined by it, so that it is not an accident that, for example, the stars radiate light energy to infinity rather than sucking it in; that, because radiation is essential to life, this then uniquely determines the direction of biological differentiation in time, and hence our psychological sense of past and future; and finally, that the thermodynamic arrow is connected in inanimate nature with the electromagnetic one, and in the laboratory with the psychological one in the manner described above. However, while it is easy to think, at each step, of reasons why the required connection *might* hold, it is probably fair to say that in no case has a connection been established with anything approaching rigour, and it would be a brave physicist who would stake his life on the assertion, for example, that conscious life must be impossible in a contracting universe. This fascinating complex of problems will probably continue to engage physicists and philosophers (and biologists and psychologists) for many years to come.

The quantum measurement paradox

If one wishes to provoke a group of normally phlegmatic physicists into a state of high animation—indeed, in some cases strong emotion—there are few tactics better guaranteed to succeed than to introduce into the conversation the topic of the foundations of quantum mechanics, and more specifically the quantum measurement problem. Apart perhaps from the old question of 'nature versus nurture' in biology or psychology, there is probably no issue in all of current science—and certainly not in the physical sciences—on which the views held by appreciable numbers of practising scientists span so wide a range, or are held so strongly.

At one extreme, there is a school of thought which holds that all the problems in this area were solved long ago, and that anyone who even thinks about them must *ipso facto* be wasting his or her time; at the other, there are those who feel that there are unresolved questions so fundamental and so urgent that there is little point in continuing to do physics in the present mould as long as they remain unsettled.

Let's go back to the classic two-slit diffraction experiment described in Chapter 1. We saw there that if we try to observe which of the two slits a given electron passes through, by setting up some detection apparatus opposite each slit, we always get a definite result—that is, each electron indeed appears to come through one slit or the other. However, under these circumstances the distribution of electrons on the second screen shows no interference effects. On the other hand, if we do *not* observe which slit the electron passes through, we get on the second screen an interference pattern characteristic of a wave which can propagate simultaneously through *both* slits. The electron appears to behave in quite different ways depending on whether or not it is being observed!

This phenomenon is a quite general feature of the quantum-mechanical formalism and its standard interpretation. Suppose we have some system which we believe is described by quantum mechanics—which means, in principle, according to the orthodox view, any physical system whatsoever—and that it has available to it several different states A, B, C, D, E, F, and can proceed between them in the way shown in Figure 5.2. The point to note

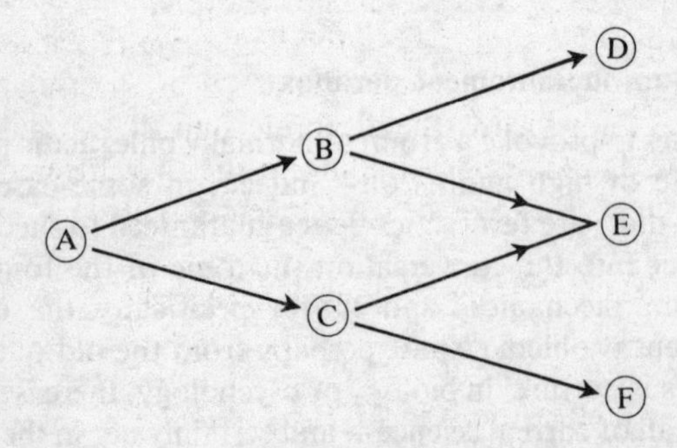

Figure 5.2 A quantum-mechanical system making transitions between possible states

is that state E can be reached from A via either B or C. Suppose now, for simplicity, that the arrows actually indicate a process in time, so that after a certain interval we know that the system has reached the B or C stage. If at this point we do an experiment to determine whether the system is in state B or state C, we will always get an unambiguous answer: it will always be found to be in state B *or* state C, never in some kind of combination of the two. Moreover, while the outcome at any particular trial will apparently be random, we will find that if we carry out a large number of trials, there are well-defined and reproducible probabilities of getting results B and C respectively. Furthermore, we can take similar statistics on the probabilities that a system observed at the intermediate stage to be in state B will proceed to the final state D or E respectively. In this way we can determine experimentally the probabilities $P_{A \to B \to D}$ ('the probability that the system proceeds from state A through state B to state D'), and so on. Note that with this experimental set-up the probability of starting from A and ending up at E is *by construction* the sum of the probabilities $P_{A \to B \to E}$ and $P_{A \to C \to E}$. (It clearly cannot be smaller, and if it were greater, we should have to conclude that we had missed one or more additional possible intermediate states.)

Now suppose we do *not* choose to observe which of the states B and C the system passed through at the intermediate stage, but rather leave the system completely isolated until the final stage. We can then measure the total probability, $P_{A \to E}$, of starting at A and arriving at E. Do we then find that this is the sum of $P_{A \to B \to E}$ and $P_{A \to C \to E}$? In general, no. $P_{A \to E}$ may be either larger or smaller than this sum, and in fact the quantitative results are compatible with the hypothesis that $P_{A \to E}$ is the square of an amplitude $A_{A \to E}$ which is the algebraic sum of two amplitudes $A_{A \to B \to E}$ and $A_{A \to C \to E}$ corresponding to the two different routes (and, of course, $P_{A \to B \to E} = (A_{A \to B \to E})^2$, etc.). Thus

$$P_{A \to E} = A^2_{A \to E} = (A_{A \to B \to E} + A_{A \to C \to E})^2 =$$
$$A^2_{A \to B \to E} + A^2_{A \to C \to E} + 2A_{A \to B \to E}A_{A \to C \to E}$$

This is not equal to the sum of $P_{A \to B \to E}$ and $P_{A \to C \to E}$, which is simply $A^2_{A \to B \to E} + A^2_{A \to C \to E}$. This is the obvious generalization of the hypothesis we had to make in the two-slit experiment (see

Chapter 1). The above results are a clear and unambiguous prediction of the formalism of quantum mechanics and demonstrate the characteristically quantum-mechanical feature of 'interference' of unobserved possibilities; for reasons we shall encounter later, the process of actually checking them experimentally for an arbitrary system is by no means trivial, but they have, in fact, been checked for a number of microscopic systems with spectacular success. For example, in the neutron diffraction device known as a neutron interferometer the states B and C can be made to correspond to states in which the neutron has not only a different spatial position (as in the simple two-slit experiment), but also a different spin orientation, and the interference between these two states is indeed seen. Similarly, a beam of neutral K-mesons has available to it two different states, called K_L and K_S, with different modes of decay. If a given meson does indeed decay during its flight through the apparatus, this in effect yields a measurement of whether it was K_L or K_S, and it always turns out to be one or the other; if, however, it does not decay, then one can see the effects of interference between the K_L and the K_S states! Another even more spectacular example will be discussed below. One can think of the 'leaving open' of, and hence interference between, unobserved possibilities on the one hand, and the fact that observation yields a definite result on the other, as generalizations, respectively, of the 'wave' and 'particle' aspects of the behaviour of an electron in the two-slit experiment.

Within the formalism of quantum mechanics in its conventional interpretation, therefore, it looks as if in some sense a system does not possess definite properties until we, as it were, force it to declare them by carrying out an appropriate measurement. But is this the only possible interpretation of the experimental data? Could it not be that the quantum formalism is merely a crude description corresponding to our present level of understanding, and that underlying it is a deeper level of reality, somewhat as the thermodynamic description of macroscopic bodies has turned out to rest on a more microscopic description—namely, that given by molecular theory and statistical mechanics? If this is the case, then it is perfectly possible that at the deeper level systems always do, in fact, have objective properties, whether or not anyone is measuring them. Moreover, such a scheme has an additional attraction: whereas, in the standard formalism of quantum mechanics, we can predict only the probability of a given outcome of a measurement, and it is forbidden to ask why on a particular trial we got the result we did, if, rather, this formalism is covering up a deeper level, then it is perfectly possible that at this level the outcome of each individual measurement is uniquely determined. The apparently random outcomes predicted by the quantum formalism would then simply be due to our ignorance of the details of the deeper-level description—just as we describe the behaviour of a macroscopic body by statistical mechanics, a probabilistic theory, even though we believe that (at least within the framework of classical physics) each event occurring in it was in fact uniquely determined by the exact history of the molecular motion (which, of course, in practice we cannot know).

Theories of the general type just described are usually called 'hidden-variable' theories; they clearly have many attractive properties, and a great deal of research has been done on them. However, it turns out that any hidden-variable theory must, if it is to agree with experiment, possess some bizarre and, to many physicists, unwelcome features. Indeed, for many years there was a general belief that *no* hidden-variable theory could be constructed which could give experimental predictions in agreement with the well-tested predictions of quantum mechanics, even for such simple systems as a particle of spin $\frac{1}{2}$ subjected to measurements of the various components of spin. This belief actually turns out to be incorrect, and for such simple cases specimen hidden-variable theories have been explicitly constructed and shown to give precise agreement with the quantum predictions, and hence with the experimental data. (Naturally, in such theories explicit account has to be taken of the way in which the carrying-out of a measurement affects the subsequent state of the system.) For certain situations involving two particles which have interacted in the past but are now separated and distant from one another, however, a hidden-variable description which gives agreement with the quantum predictions can be maintained only at the cost of sacrificing one or more basic assumptions about the world which most people would dearly like to keep. The proof of this remarkable theorem, which is originally due to J. S. Bell in 1964 and has since been refined by many other people, is so simple as to be worth giving here.

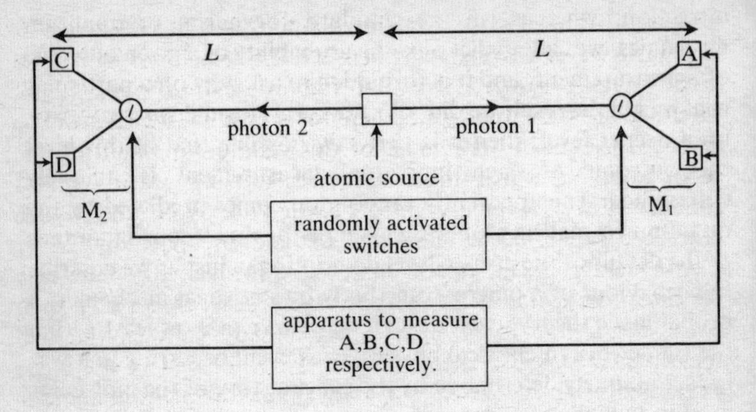

Figure 5.3 Schematic set-up of an experiment to measure the polarization of photons emitted in an atomic cascade process

For the sake of concreteness let us focus our attention on the system which is actually used in most of the relevant experiments: namely, a set of identical atoms which can be excited from their lowest energy state and can then decay by a 'cascade' process—that is, a process which involves two transitions, from the initial excited state to an intermediate state and then from the intermediate state back to the final (lowest) state. At each of the two stages of the cascade a photon is emitted, and it is the pairs of photons which are our real object of study; the role of the atoms is merely to provide the photons. Let us restrict attention to those pairs in which one photon is emitted (say) along the $+z$ axis and the other along the $-z$ axis; moreover, let us assume that we have a means of identifying unambiguously the correct 'partner' of a given photon—that is, the one which was emitted by the same atom. (This could be done in principle by making the gas of atoms so dilute that the average time interval between cascade processes in different atoms is long compared to the lifetime of the intermediate atomic state. In practice the situation is a bit more complicated, but this is not a real difficulty.) We arrange to make polarization measurements on the two photons, in a way described below, at points some distance away in the $+z$ and the $-z$ directions respectively (see Figure 5.3). Suppose that the distance from the source to each apparatus is L. If the first photon is emitted at time zero, say in the $+z$ direction, it will arrive at apparatus M_1 at a time L/c, where c is the velocity of light. The atom, having emitted the first photon and dropped to its intermediate state, then waits there for a time Δt, which is of the order of the mean lifetime, τ, of this state before emitting the second photon, which therefore arrives at its apparatus M_2 at a time $\Delta t + L/c$. The measurements carried out by M_1 and M_2 are therefore separated in time by Δt and in space by $2L/c$. Consequently, if L is greater than $c\Delta t/2$ (which will be the case for most pairs, provided L is appreciably larger than $c\tau/2$), then there is no time for a light signal to pass between the two measurement events; and consequently, if we believe the basic principles of special relativity, they cannot affect one another causally. In the most spectacular series of such experiments to date, L was about 6 metres and the lifetime of the intermediate state was about 5×10^{-9} seconds, so the criterion was well satisfied.

The measurements we wish to perform on the photon pairs are actually what would normally be called measurements of linear polarization. For present purposes it is unnecessary to go into the details, except to remark that quantum mechanics makes unambiguous predictions about the outcome of such measurements: it is sufficient to schematize each measuring apparatus as simply a black box with a single adjustable control parameter (call it p_1 for M_1 and p_2 for M_2), which for each setting is guaranteed to yield, for each incident photon, one of two answers, say yes or no. Let us consider a particular setting, a, of the control parameter p_1 of apparatus M_1. We define a variable A such that if a given photon is incident on M_1 and the answer given is yes, then this photon is said to have been measured to have $A = +1$; if no, then $A = -1$. Similarly, if the setting of the control parameter p_1 is b, we define a variable B such that for a given photon, B is measured to be $+1$ or -1 according as the answer given by M_1 is yes or no. Note that if the setting of p_1 is b, then A is (so far) undefined, and vice versa. Similarly we consider settings c and d of the control parameter p_2 of the apparatus M_2, and define variables C and D which can take the values $+1$ and -1 in the same way. For any given photon 1, therefore, we can measure *one* of the quantities A and B (but not both), and similarly for any given photon 2, one of the quantities

C and *D*. Hence for any given *pair* we can measure one and only one of the product quantities *AC*, *AD*, *BC*, and *BD*, and in each case by construction the answer will be +1 or −1. The experiment consists in taking measurements on a large number of pairs, with the setting of M_1 alternated between a and b, and that of M_2 between c and d, so that we have a large number of measurements of each of the quantities *AC*, *AD*, *BC*, and *BD*. The basic experimental data are the average values of these quantities, which we will denote by $\langle AC \rangle$ and so on.

Before embarking on a discussion of what we should expect to find in this experiment, I should add an important note. We will assume that the control parameters p_1 and p_2 can be arbitrarily set 'at the last moment'—that is, just before the photon in question enters the apparatus. The significance of this is that then not only can the behaviour of photon 2 not be causally affected by the *outcome* of the measurement made on photon 1, but it also cannot be causally affected by our decision about which quantity (*A* or *B*) we are going to measure. Actually, to change the value of the control parameter for a given apparatus at such short notice—namely, in less than about 10^{-9} seconds—is technically difficult, but in the most spectacular version of the experiment (by Alain Aspect and his colleagues in Paris) an equivalent arrangement (or so one hopes) is established by having two different apparatuses set up for photon 1, with pre-set values a and b respectively, and diverting the photon into one or the other by means of a randomly activated switch; and likewise for photon 2 (see Figure 5.3).

Let us now make the following apparently innocuous set of assumptions. First, each photon 1 possesses a definite value of the variable *A*, whether or not this quantity is measured; similarly of *B*. Likewise, each photon 2 possesses definite values of *C* and *D*, irrespective of which of these quantities, if either, is measured. Second, the values of *C* and *D* possessed by photon 2 cannot be affected by whether it is *A* or *B* which the distant apparatus M_1 measures for photon 1 (or by the outcome of the measurement); similarly, the values of *A* and *B* cannot be affected by the setting of M_2. Third, the average value of $\langle AC \rangle$ obtained for the pairs on which we actually measure this combination is identical (with the usual provisos regarding statistical fluctuations) with the average value for all the pairs; that is, the sample of pairs on which we happened to measure AC is typical of the whole set, and similarly for $\langle BC \rangle$ and the others. A theory which satisfies these three assumptions, or slight variants of them, is called an 'objective local theory'. This class of theories is slightly more general than that of local hidden-variable theories, which it contains as a subclass.

It is worth stressing just how innocuous our assumptions apparently are. The first assumption, for example, does not imply that we necessarily have, in principle, a method of knowing A without actually measuring it, or that the value of A would be unaffected by the measurement process; all it really implies is that for each photon 1 there is a definite result which we 'would have' got had we measured A. The second assumption, given the geometry of the experiment, is (at first sight at least!) simply an application of one of the basic principles of special relativity; since the events of setting M_1 and making the measurement with M_2 have spacelike separation, they cannot affect one another causally.[3] Finally, the third assumption is at first sight simply an application of the usual principle of induction, which says that if I measure a quantity on a subset chosen at random from a given set, the results I get should be typical of the set as a whole.

Let us now do a very simple piece of algebra. By the first assumption, each photon 1 possesses a value ± 1 of A and also a value ± 1 of B, which by the second assumption is independent of whether C or D is measured. Similarly, each photon 2 has a value ± 1 of C and ± 1 of D which is independent of whether A or B is measured. Consequently, each photon pair has a value ± 1 of each of the quantities AC, BC, AD, and BD, and hence of the combination $(AC+BC+AD-BD)$. Now one can easily convince oneself that for each of the 16 different cases corresponding to the possible choices ± 1 for each of A, B, C, D separately, this combination is always either $+2$ or -2; hence its average over the whole set of pairs cannot possibly be greater than 2 (or less than -2). Since the average of a sum of terms is equal to the sum of the averages, this implies that

$$\langle AC \rangle + \langle BC \rangle + \langle AD \rangle - \langle BD \rangle \leqslant 2$$

where the averages refer to the average values for the whole set of pairs.

Finally, by virtue of the third assumption we can identify the averages $\langle AC \rangle$ and so on with the experimentally measured values of this quantity. Thus we have a clear prediction concerning the *actual* measured quantities in this experiment.

Now comes the crunch. Quantum mechanics makes quite clear and unambiguous predictions concerning the values of the quantities $\langle AC \rangle$ and the rest—and, with suitably chosen values of the settings, they do *not* satisfy the above inequality! Moreover, the outcome of the experiment is quite unambiguous: the data agree with the quantum-mechanical predictions within the experimental error, and quite plainly violate the above inequality. Thus, at first sight at least, we are forced to sacrifice at least one of the three assumptions which define an objective local theory.

This conclusion is so surprising—indeed, to some, alarming—that strenuous attempts have been made over the last twenty years to find loopholes in the theoretical argument or in the analysis of the experiments. Such loopholes can indeed be found, but however many have been closed (as some have been, and others probably will be, by improvements in experimental technique), a sufficiently ingenious objector will almost certainly find yet more (in the last resort, there is no *logical* rebuttal to the argument that all the experimental data assembled to date may be a giant statistical fluke!). All one can say is that most of these objections seem to most physicists so contrived and *ad hoc* that in any other context they would be dismissed out of hand. Whether one believes that the a priori arguments in favour of local objectivity are so compelling that it is legitimate to grasp even at such straws to save it must of course remain a matter of taste.

If one decides to ignore the loopholes, which of the three ingredients of an objective local theory is the least painful to give up? Most people would be extremely reluctant to sacrifice the second, the principle of local causality; for at first sight, at least, the whole framework of quantum field theory would then crumble. Again, the idea of abandoning the third—that is, the assumption that the particular pairs on which we actually carry out a measurement of a particular property are characteristic with respect to that property of the set of pairs as a whole—seems very repugnant to most people. It is perhaps worth pointing out, however, that this repugnance is connected with our deeply ingrained belief that the state of a system (in this case the photon pair) in the time interval between emission and measurement is a function of the *initial* conditions (that is, the state of the emitting atoms), not of the *final* conditions (that is, the result of our measurement). It is conceivable that in a new world-view in which our ideas about the arrow of time were radically revised we would be less unhappy at the prospect of abandoning the third assumption. Be that as it may, the majority view at present among physicists is that the least of the possible evils is to abandon the first assumption, and to agree with Niels Bohr that microscopic entities such as photons cannot even be conceived of as possessing definite properties in the absence of a measuring device set up to determine those properties. This assertion is one of the main corner-stones of the so-called 'Copenhagen interpretation' of quantum mechanics, of which Bohr was a leading exponent, and which could fairly be called the dominant orthodoxy of the last sixty years. To ask how a photon 'would have' behaved if subjected to a different measuring apparatus is in this view totally meaningless. What the atomic cascade experiments and their interpretation illustrate particularly dramatically is that, contrary to a misconception which probably arose from some influential early popular writings of Heisenberg and dominated discussion of the issue for at least the next decade, this fundamental feature of quantum mechanics has nothing whatever to do with any possible physical disturbance of the system by the measuring apparatus; in the analysis of these experiments, given local causality, there is simply nowhere that any 'disturbance by measurement' could play a role.

If this were all—if microscopic entities were simply different in their ontological status from the familiar macroscopic ones, as Bohr so often seems to imply in his writings—then we could all rest assured that the situation, while surprising, poses no fundamental philosophical difficulties. (In the thirties and forties, a number of professional philosophers did indeed come to terms with quantum mechanics along the general lines laid down by Bohr.) But alas, worse is to come. Indeed, the really vexing problem in the foundations of quantum mechanics is not that there is a profound difference between the microscopic world of electrons and photons and the macroscopic world of chairs, tables, Geiger counters, and photographic plates, but that there is no such difference. Quantum mechanics claims to be a complete and unified description of the world, and as a consequence, if it possesses counter-intuitive or alarming features at the micro-level, there is no way to prevent them from propagating upwards and eventually infecting also the macro-level. To be precise, let us examine the notion of 'measurement'. Suppose, for example, that we wish to 'measure' which slit the electron passed through in the standard two-slit experiment. We might do this by placing some kind of particle counter—a Geiger counter or something similar—opposite each of the slits; call the counters C_1 and C_2 respectively. Then either C_1 will register the passage of the electron and C_2 will not, or vice versa: clearly the electron has passed through one slit or the other, not both. How do we know this? Because we actually hear one and only one counter click (or we have the result recorded on a computer print-out which we subsequently inspect, or whatever); essentially, it is a matter of direct experience.

But wait. The counter is not some kind of magic device beyond the laws of physics; it is simply an arrangement of gas molecules, electrodes, and so on set up in a particular way, and as such should surely be subject to the usual laws of physics, and in particular to the principles of quantum mechanics. This means, among other things, that as long as it is unobserved it should behave in a wave-like manner—that is, that its state should continue to represent all possibilities. In particular, application of the quantum formalism to the electron *plus* counters leads inescapably to a description in which it is quite incorrect to say either that counter C_1 has fired and counter C_2 has not, or vice versa; rather, both possibilities are still present. This is not an accidental feature which could be removed, for example, by modifying our detailed hypotheses about the way the counters work; it is built deep into the foundations of the quantum formalism. At least until we reach the level of the conscious human observer, there is simply no point at which the formalism allows the introduction, other than by simple fiat, of the concept of 'measurement' interpreted as a process which is guaranteed to select a single definite result from among a number of possibilities. Thus, the lack of objectivity which appears to characterize the quantum description at the micro-level propagates up to the macro-level. This disturbing feature is illustrated by the famous 'Schrödinger's cat paradox', which in slightly modified form goes as follows. We add to the apparatus described a closed and sound-proof box containing a cat and a means of executing the latter, and arrange by appropriate electronics and so on that if the counter C_1 fires, the cat is killed, whereas if C_2 fires, nothing happens to it. Then, given a suitable choice of initial state of the electron, the quantum formalism predicts unambiguously that after the electron has passed the screen with the slits, the cat will be described by a state which corresponds neither to its being certainly alive nor certainly dead, but has both possibilities still represented in it! On the other hand, it is a matter of common sense (and can be easily checked, if not with cats, then at least with counters) that if we open the box and inspect the cat, we will find it either alive or dead, not in some strange combination of the two.

We can summarize the situation succinctly by saying that in the quantum formalism things do not 'happen', while in everyday experience they do. This is the quantum measurement paradox. As indicated above, it is regarded by some physicists as a non-problem and by others as undermining the whole conceptual basis of the subject. Let me try to describe briefly some of the most popular 'resolutions' of the paradox.

Among those working physicists who regard the problem as trivial, the most popular of the alleged resolutions is probably that which runs, schematically, as follows. It is first pointed out that the only reason we have for believing that the electron, when not observed, in some sense propagated through both slits, or more generally, in the case discussed earlier in this chapter, that the system passed through both the states B and C, is that we can observe experimentally the effects of the interference between the two possibilities. Next, it is remarked that any measuring device must necessarily be a macroscopic system, and, moreover, must be constructed in such a way that the different possible states of the microscopic system (electron or photon) must induce *macroscopically different* states of the apparatus (thus, if the electron passes through slit 1, counter C_1 fires, otherwise it does not, and the states of being fired or not are macroscopically distinguishable). Further, in order to protect the result of the measurement against thermodynamic fluctuations, the device must have built into its working a substantial degree of irreversibility. The next step is to observe that a careful application of the quantum formalism to a system having these features leads to the conclusion that the effects of interference between states which are macroscopically distinct are unobservable—in practice, if not in principle—and therefore that all experimental predictions concerning the behaviour of the device will be precisely the same as if it were indeed in one macroscopic state or the other (but we do not know which). Finally, the 'as if' is slyly dropped, and one argues that 'therefore' the device actually *is* in one macroscopic state or the other (and that Schrödinger's cat actually *is* alive or dead).

All steps but the last in this argument are generally agreed to be unproblematic. (With regard to the claim that macroscopic systems embodying a normal degree of irreversibility cannot show interference between macroscopically distinct states, there must by now be literally hundreds of papers in the literature which illustrate this point in more or less realistic examples.) What of the last step? Many physicists find it quite acceptable; and indeed, if one takes the point of view that two alternative descriptions of a system which make identical predictions for the outcome of all possible experiments must *ipso facto* be equivalent, then there is no problem. A minority (which includes the present author) take the view that there is an unbridgeable logical chasm between the statements that the system behaves 'as if' it were in a definite macroscopic state and that it *is* in such a state, and therefore regards the last step in the argument as the intellectual equivalent of an illusionist's trick, which totally fails to solve the real problem.

If one is unsatisfied with the above resolution, what is left? One logically consistent possibility is to apply to the quantum description of macroscopic systems the same interpretation as is often applied at the micro-level—that is, to see the quantum description as nothing but a formal calculus whose sole purpose is to predict the probability of various macroscopic events (for example, the (audible) clicking or not of a particular counter). On this view, questions like 'What was the state of Schrödinger's cat before we inspected it?' are simply not within the competence of quantum mechanics to answer, and should therefore not be asked. (Just as it is often claimed that, at the micro-level, questions like 'Which slit did the electron actually go through?' are similarly meaningless.) In its extreme form, such a view might claim that such questions will be for ever unanswerable, and hence perhaps meaningless; in a milder form, it would leave open the theoretical possibility that they may someday be answerable by a theory which supersedes quantum mechanics. In either case, however, this view conflicts rather violently with our everyday assumption that questions about the macroscopic state of a macroscopic object are not only legitimate but have in principle definite answers. If eventually we should find we have to live with it, it would probably be at the cost of considerable mental discomfort.

Various other resolutions of the paradox have been proposed—for example, that human consciousness plays an essential role, but one which is not describable in terms of the laws of physics. Since our current understanding of the phenomenon of consciousness is probably even less adequate than our understanding of the foundations of quantum mechanics, I do not feel it would be particularly profitable to discuss this or similar proposals here. However, there is one so to say exotic solution that needs to be mentioned, if only because it has been widely advertised in recent popular books on quantum mechanics and cosmology: namely, the interpretation variously known as the Everett–Wheeler, relative state, or many-worlds interpretation. The basis for this alleged solution is a series of formal theorems in quantum measurement theory which guarantee that the probability of two different 'measurements' of the same property yielding different results (of course, under appropriately specified conditions) is zero. Crudely speaking, the probability of my hearing counter 1 click at the passage of a particular electron while you hear counter 2 click is zero. The 'many-worlds' interpretation, at least as presented by its more enthusiastic advocates, then claims, again crudely speaking, that our impression that we get a particular result on each experiment is an illusion; that the alternative possible outcomes continue to exist as 'parallel worlds', but that we are guaranteed, through the above formal theorems, never to be conscious of more than one 'world', which is, moreover, guaranteed to be the same for all observers. The non-observed worlds are said to be 'equally real' with the one we are conscious of experiencing. The more extreme adherents of this interpretation have drawn various conclusions for cosmology, psychology, and other disciplines which to the non-believer must seem distinctly exotic.

Let me allow myself at this point the luxury of expressing a strong personal opinion, which would certainly be contested vigorously by some of my colleagues. It seems to me that the many-worlds interpretation is nothing more than a verbal placebo, which gives the superficial impression of solving the problem at the cost of totally devaluing the concepts central to it, in particular, the concept of 'reality'. When it is said that the 'other worlds' of which we are not, and in principle never could be, conscious, are 'equally real', it seems to me that the words have become uprooted from the context which defines their meaning and have been allowed to float freely in interstellar space, so to speak, devoid of any point of reference, thereby becoming, in my view, quite literally meaningless. I believe that our descendants two hundred years from now will have difficulty understanding how a distinguished group of scientists of the late twentieth century, albeit still a minority, could ever for a moment have embraced a solution which is such manifest philosophical nonsense.

The reader will probably have gathered by now that I do not myself feel that *any* of the so-called solutions of the quantum measurement paradox currently on offer is in any way satisfactory. This is true, and my dissatisfaction is the point of departure for various speculative remarks I will make in the concluding chapter.

6
Outlook

One of the luxuries allowed to the author of a book such as this is the opportunity to attempt, however amateurishly, to forecast some of the major directions in which his subject will move in the next few decades. It is a dangerous undertaking. The history of physics is littered with confident predictions, often by the leading minds of the period, which even a few years later have looked distinctly short-sighted. Nevertheless, the temptation to speculate, or at least to throw out a few questions, is irresistible, and I shall succumb to it. It goes without saying that the opinions expressed in this chapter are personal ones, and are in no way intended to reflect the consensus—if indeed one exists—of the physics community as a whole; indeed, if anything, they are probably violently opposed to the viewpoint of the majority.

A few years ago Stephen Hawking, a leading cosmologist and particle physicist, gave a lecture entitled 'Is the end in sight for theoretical physics?' and proceeded to answer the question with a qualified yes. His thesis was that we should probably soon attain a unification not only of the strong interaction with the electroweak one—that is, a 'grand unified' theory (see Chapter 2)—but of both with gravity; that the data of cosmology would similarly be explained; and, in effect, that there would then be no interesting questions left. Such confident predictions have become increasingly common over the last decade or so, particularly with the advent of the so-called superstring theories, with their promise of possibly predicting all particle masses and interactions without any arbitrary input parameters; indeed, one frequently hears such theories colloquially referred to as actual or potential 'theories of the world', or 'theories of everything'.

One assumption implicit in such a view, of course, is that if we can understand the data of particle physics and cosmology, then we can 'in principle' understand everything about the world; or, as one recent popular article succinctly phrased it, that 'the four basic interactions . . . together with cosmology, account for all known natural phenomena'. I will return below to the status of this assertion; since it is probably believed by a substantial majority of physicists (particularly particle physicists and cosmologists!), let us temporarily grant it. Are we then indeed nearly at the end of the road?

An outsider viewing the progress of particle physics and cosmology over the last few decades might well compare its practitioners, I think, to a mountaineering party crossing a badly crevassed glacier. As each new chasm looms ahead, it is bridged or outflanked by some ingenious manœuvre, each more daring and spectacular than the last; and devious and tortuous as the route seems, the party moves slowly but steadily forwards. Is the short range of the weak interaction a problem? Then postulate a heavy weak intermediate boson to mediate it. Does the mass of this particle lead to apparently insuperable technical problems? They are overcome by the Higgs mechanism. Do the symmetries postulated in particle physics produce unwanted cosmological consequences? Perhaps they can be washed out of sight by a period of 'inflationary' expansion of the universe. And so on. All the while the party is relying on a single trusted guidebook—the guide provided by the conceptual structure of quantum field theory, regarded by a majority, at least, of physicists as the ultimate truth about the physical universe.

Needless to say, the manœuvres performed by the party are not arbitrary or *ad hoc*; most of them solve a number of difficulties simultaneously, and in some cases, at least, their predictions have been spectacularly confirmed by experiment, as, for example, in the discovery of the intermediate vector bosons. But what must seem truly amazing, if we lift our eyes a little from the day-to-day route-finding problems, is the way in which these ingenious developments have succeeded for nearly sixty years in keeping our understanding of the physical world confined (the word is not meant pejoratively) within the bounds set by quantum field theory. Indeed, despite the frequent use of the phrase 'the new physics' to refer to the developments of the last two decades, the basic conceptual framework within which physicists interpret the world today, if viewed against the background of three hundred years of history, is really very little different from what it was in 1930; even the recent superstring theories, which replace the concept of point particles by one-dimensional (string-like) entities, really change only the players, not the rules of the game.

Will the quest for mathematical consistency within the framework of quantum field theory continue to guide us between the crevasses in the future as it has in the past, and maybe eventually place us safely on the far side? Will we find eventually that its stringent constraints, whether in the superstring version or some other, not only permit but mandate the world we know? Or will we eventually discover, as more and more consequences are worked out, that there is not one consistent solution within the framework set by the constraints, but none? Could it be—as a cynic might suggest—that the only reason that we seem, in 1987, to have a reasonably consistent picture of the structure of the universe is that we do not have the technology to do the experiments which would prove it wrong? After all, many modern theories involve speculation about the behaviour of matter at energies which lie even further beyond the reach of current or projected accelerators than the latter do beyond the characteristic energies of everyday life. If Nature has some surprises in store for us, she certainly has plenty of room in which to spring them!

A rather different question is: Even supposing that quantum field theory does succeed in explaining the structure of all the basic interactions and so on, will we be content with it as a final 'theory of the universe'? My own guess is that we will not. As sociologists and historians of science have often observed, dissatisfaction with the basic conceptual structure of a scientific theory tends to be suppressed as long as that theory is turning out answers at a satisfactory rate, but to re-emerge the moment it fails to do so—and that may well be the case even if the reason is that there are no more questions! I suspect that within a generation or two we will begin to feel, no doubt subconsciously at first, the need to formulate new kinds of questions which cannot even be asked within the framework of quantum field theory. But the motivation for such questions may well come from a direction quite different from particle physics, and I will return to this below.

Let's suppose now, for the sake of argument, that we succeed in finding a complete and consistent explanation of all the data of particle physics and cosmology. Does that mean that physics, or even theoretical physics, will indeed grind to a halt? Only by a very peculiar definition of the subject. As pointed out in Chapter 4, even if the behaviour of condensed-matter systems is 'in principle' simply a consequence of that of the 'elementary' particles composing them, the task of explaining and predicting this behaviour is by no means the relatively trivial problem it is sometimes assumed to be. The non-scientific public, unfortunately, often seems to be under the impression that as long as physicists are dealing with matter under relatively 'terrestrial' conditions—as distinct, say, from those attained in high-energy accelerators or distant regions of the universe—they can always predict with complete confidence how it will behave. This is a myth. As mentioned in Chapter 4, our success in predicting the existence and properties of qualitatively new phases of matter has been limited in the extreme; and even at this moment there is at least one well-verified phenomenon in low-temperature physics which not only was not predicted in advance, but which most physicists, if given the relevant data, would have probably predicted with considerable confidence could not conceivably occur. (We still do not know why it does!) For a physicist to claim that he 'knows' with certainty exactly how a particular physical system will behave under conditions very far from those under which it has been tested—for example, the conditions under which certain types of space-based defence systems might have to perform if they were ever used—seems to me arrogant, and indeed in some contexts dangerous. Nature, alas, has a habit of thinking of clever possibilities that have not occurred to us, and we ignore her subtlety at our peril. Unless we take the view that anything other than the study of the elementary constituents of matter is 'derivative', and hence perhaps 'not really physics'—a view against which I have argued in Chapter 4—there is plenty of physics left to do in the area of condensed matter (and, of course, in many other areas of the subject which I have had no space even to mention in this book).

Let me now return to the claim that 'the four basic interactions . . . together with cosmology, account for all known natural phenomena'. This is, of course, not a statement of fact, but an act of faith. It is not an unreasonable one. What anyone who makes it is saying, in effect, is that while there are many natural phenomena which currently have *not* actually been explained in detail in terms of the four basic interactions, there is no clear example at present of a single phenomenon which we can prove cannot be so explained; so that the principle of Occam's razor suggests that we should try to get along with the interactions we know about. Of course, such a view may commit one to a particular selection of the 'facts' to be explained; for example, I strongly suspect that if the so-called 'fifth force' recently proposed (see Chapter 2) should turn out to be genuine, then we shall rapidly discover a number of phenomena (just conceivably, but improbably, including some now labelled 'paranormal') which can be naturally explained in terms of it, but which have in the past been dismissed as the result of experimental error, fraud, or hallucination. What constitutes a 'fact' in physics is not totally independent of one's theoretical preconceptions!

A question which I personally find even more fascinating is this: Is the behaviour of complex systems indeed simply a consequence of the 'complex interplay among many atoms, about which Heisenberg and his friends taught us all we need to know long ago'—even in the rather weak sense considered in Chapter 4? Or does the mere presence of complexity or organization or some related quality introduce new physical laws? To put it another way, would the complete solution of the basic equation of quantum mechanics—Schrödinger's equation—for, say, the 10^{16}-odd nuclei and electrons composing a small biological organism actually give us, were it achievable, a complete description of the physical behaviour of such an organism? The conventional answer to the question is undoubtedly yes. But what few people, outside or indeed within physics, seem to realize is the flimsiness—or rather, the complete absence—of positive experimental evidence for this conclusion. It is indeed true that application of the formalism of quantum mechanics to complex systems yields predictions for currently measurable quantities which are often in good agreement, quantitatively as well as qualitatively, with the experimental results; and that in cases where there is substantial disagreement, there are usually enough unknown factors in the experimental system or approximations in the theory that the discrepancy can be plausibly blamed on one or other or a combination thereof. Certainly there is at present no clear evidence that such quantum-mechanical calculations give the wrong answers. What is rarely appreciated, however, is that, in the context of meaningful tests of quantum mechanics, in almost all cases up till now one has been dealing with very 'crude' features, which are in some sense the sum of one-particle (or 'one-quasi-particle') properties, or at best properties of pairs of particles. Where the *specifically quantum-mechanical* aspects of the behaviour of a complex system can indeed be regarded as effectively the sum of the contributions of such small groups of microscopic entities, quantum mechanics seems to work well. Beyond this, despite everything that was said in Chapter 4,[1] it has barely begun to be genuinely tested.

At this point the reader may well exclaim impatiently, 'But you admit that quantum mechanics works well—spectacularly well, in fact—for single atoms and molecules, and even small groups of atoms and molecules. Since complex bodies are composed of atoms and molecules, isn't it *obvious* that quantum mechanics must work equally well for complex bodies? Surely that is common sense?' Indeed it is—except that the history of physics has shown us repeatedly that 'common sense' can be wrong! Certainly, the reductionist principle—put crudely, that the whole is no more than the sum of its parts—has served us well in the past; though I suspect that our affection for it may also have a certain anthropomorphic element about it, in that we tend subconsciously to rely on our experience of taking apart things put together by other human beings to 'see how they work' and then automatically to assume that Nature works on the same principles. Be that as it may, the principle of Occam's razor certainly favours reductionism, and the only obvious reason to query it—to me anyway, though certainly not at present to a majority of physicists, a powerful one—is the quantum measurement paradox discussed in Chapter 5, which suggests the question: Does the formalism of quantum mechanics still apply even when it appears to suggest that a macroscopic object need not be in a definite macroscopic state?

In the last few years, various experiments have been started which have as part of their goal an exploration of whether, and if so where and how, quantum mechanics breaks down in the face of increasing complexity; at the level explored so far, no evidence for any breakdown has been found. Whatever the outcome of this particular experimental programme (and the work mentioned at the end of Chapter 4, which, although somewhat differently oriented, may well turn out in the end to be relevant to these issues), I am personally convinced that the problem of making a consistent and philosophically acceptable 'join' between the quantum formalism which has been so spectacularly successful at the atomic and subatomic level and the 'realistic' classical concepts we employ in everyday life can have no solution within our current conceptual framework; that it is only a matter of time before the increasing sophistication of the technology available makes this not just a philosophically, but an experimentally, urgent question; that the resulting conceptual impasse will eventually generate a quite new description of the world, whose nature we cannot at present even begin to guess; and that at some date in the future—fifty years, a hundred years, from now?—quantum field theory and indeed the whole quantum-mechanical world-view will appear to our descendants, as classical physics does to us, as simply an approximate description which happened to give the right answers for the types of experiment thought feasible and interesting by physicists of the late twentieth century.

If even a small part of the above speculation is right, then, far from the end of the road being in sight, we are still, after three hundred years, only at the beginning of a long journey along a path whose twists and turns promise to reveal vistas which at present are beyond our wildest imagination. Personally, I see this as not a pessimistic, but a highly optimistic, conclusion. In intellectual endeavour, if nowhere else, it is surely better to travel hopefully than to arrive, and I would like to think that the generation of students now embarking on a career in physics, and their children and their children's children, will grapple with questions at least as intriguing and fundamental as those which fascinate us today—questions which, in all probability, their twentieth-century predecessors did not even have the language to pose.

Notes

Chapter 1: Setting the stage

1. The angular momentum of a molecule relative to a given reference point is its momentum times the perpendicular distance of its path from the point. The total angular momentum is this quantity summed over all the molecules.
2. An amusing historical footnote: it seems likely that Planck himself would have preferred that history should have attached his name not to h but to the much less fundamental k_B.
3. The original German word *Unbestimmtheitsprinzip* ('indefiniteness' or 'indeterminacy principle') is sometimes mistranslated 'uncertainty principle'. This is very misleading, since it suggests that the electron actually has a definite position and momentum of which one is uncertain. In fact, the quantum formalism simply does not allow the ascription of a definite position and momentum simultaneously.

Chapter 2: What are things made of?

1. Ironically in the present context, the word 'atom' means literally 'indivisible'.
2. The lifetime is about 1.4 times the 'half-life'—that is, the time after which approximately half the original sample is left.
3. It is interesting that this principle, which most contemporary physicists probably regard as totally beyond question, has not always seemed so. During the birth-throes of quantum mechanics, a famous paper by Bohr, Kramers, and Slater raised the question of whether energy conservation applied to individual processes or was true only on the statistical average. Experiment soon gave the first answer.
4. Actually, it is at present an open question whether the neutrino really has zero rest mass or just a very small one (perhaps around 10 eV).
5. For completeness it should be added that in a very few nuclei, for reasons which go somewhat beyond the scope of the present discussion, *both* neutron and proton are unstable.

6. In the case of massless particles, which cannot be brought to rest in any reference frame, a slightly different definition of the spin is necessary.
7. Compare the discussion of the electron in the atom, given on pp. 25–6.
8. The word 'gauge' was attached for purely historical reasons; the term sometimes used in the Russian literature—'gradient invariance'—is far more informative.

Chapter 3: The universe: its structure and evolution

1. W. Rindler, *Essential Relativity*, van Nostrand Reinhold, New York, 1969, p. 213.
2. The element helium was actually first detected spectroscopically in the sun, whence its name.
3. In a space of (local) positive curvature the surface area of a sphere is less than 4π times the square of its radius (just as the circumference of a circle in the corresponding two-dimensional case was less than 2π times its radius); in a space of negative curvature it is greater.

Chapter 4: Physics on a human scale

1. S. L. Glashow in *Physics Today*, Feb. 1986, p. 11.
2. As householders in a cold climate know to their cost, the density of the liquid phase of water is actually slightly greater than that of the solid phase (ice).
3. This does not take into account those metals, so-called superconductors (discussed later in Chapter 4), which show effectively infinite conductivity below a certain temperature.
4. With an eye to some remarks I shall make in the final chapter, it should be emphasized that the effects in question are purely classical and are not usually thought to involve subtle quantum-mechanical many-particle correlations.

Chapter 5: Skeletons in the cupboard

1. See J. D. Barrow and F. J. Tipler, *The Anthropic Cosmological Principle*, Oxford University Press, Oxford, 1986.
2. It should be mentioned that while this view is probably held by an overwhelming majority of physicists, it is not held universally. There is a school of thought associated with the

Belgian scientist Ilya Prigogine which holds that dissipation is itself a fundamental phenomenon, which cannot be analysed in the way sketched here.

3. There are actually some subtle questions at this point connected with exactly what we regard as the 'measurement' but with the most naïve interpretation the statement is true.

Chapter 6: Outlook

1. Even the phenomena of superfluidity and superconductivity, spectacular as they are, are still in the relevant sense the sum of a large number of one- or two-particle effects.

Further reading

Chapter 1

H. Butterfield, *The Origins of Modern Science 1300–1800*, Bell and Sons, London, 1950.

T. S. Kuhn, *The Structure of Scientific Revolutions*, University of Chicago Press, Chicago, 1962.

Chapter 2

S. Weinberg, *The Discovery of Subatomic Particles*, Scientific American Library, New York, 1983.

Y. Ne'eman and Y. Kirsh, *The Particle Hunters*, Cambridge University Press, Cambridge, 1986.

P. Watkins, *Story of the W and Z*, Cambridge University Press, Cambridge, 1986.

A. Pickering, *Constructing Quarks: A Sociological History of Particle Physics*, Edinburgh University Press, Edinburgh, 1984.

Chapter 3

D. Sciama, *The Physical Foundations of General Relativity*, Doubleday, Garden City, New York, 1969.

S. Weinberg, *The First Three Minutes: A Modern View of the Origins of the Universe*, Basic Books, New York, 1977.

J. D. Barrow and J. D. Silk, *The Left Hand of Creation*, Counterpoint, London, 1983.

Chapter 4

Articles by A. Bruce and D. J. Wallace, J. Ford, G. Nicolis, D. J. Thouless, and the present author in 'The New Physics' (see below).

Chapter 5

J. D. Barrow and F. J. Tipler, *The Anthropic Cosmological Principle*, Oxford University Press, Oxford, 1986.

H. Reichenbach, *The Direction of Time*, University of California Press, Berkeley, 1956.

N. Herbert, *Quantum Reality: Beyond the New Physics*, Anchor Press/Doubleday, New York, 1985.

A. Rae, *Quantum Physics: Illusion or Reality?* Cambridge University Press, Cambridge, 1986.

Note: An up-to-date survey of many frontier areas of modern physics, at a level similar to or somewhat higher than that of this book and in considerably more detail, may be found in the multi-author collection 'The New Physics', ed. P. C. W. Davies, Cambridge University Press, to be published in 1988.

Index